SpringerBriefs in Statistics

T0213905

For further volumes:
http://www.springer.com/series/8921

S. Ejaz Ahmed

Penalty, Shrinkage and Pretest Strategies

Variable Selection and Estimation

Springer

S. Ejaz Ahmed
Brock University
St. Catharines, ON
Canada

ISSN 2191-544X ISSN 2191-5458 (electronic)
ISBN 978-3-319-03148-4 ISBN 978-3-319-03149-1 (eBook)
DOI 10.1007/978-3-319-03149-1
Springer Cham Heidelberg New York Dordrecht London

Library of Congress Control Number: 2013954567

Mathematics Subject Classification (2010): 62J05, 62J07, 62F10, 62F12

Printed on acid-free paper

Springer is part of Springer Science+Business Media (www.springer.com)

Preface

This book is intended to provide a collection of topics outlining pretest and Stein-type shrinkage estimation techniques in a variety of regression modeling problems. Since the inception of the shrinkage estimation, there has been much progress in developing improved estimation strategies both in terms of theoretical developments and their applications in solving real-life problems.

Recently, LASSO and related penalty-type estimation have become popular in problems related to variable selection and predictive modeling. In this book, selected penalty estimation techniques have been compared with the full model, submodel, pretest, and shrinkage estimators in some regression models. Further, one chapter is dedicated to estimation problem in pooling data from several sources. Several real data examples have been presented along with Monte Carlo simulations to appraise the performance of the estimators in real settings. The book is suitable as reference book for a graduate course in regression analysis and combining data from several sources. The selection of topics and the coverage will be equally useful for the researchers and practitioners in this field.

This book is organized into six chapters. The chapters are standalone so that anyone interested in a particular topic or area of application may read that specific chapter. Those new to this area may read the first two chapters and then skip to the topic of their interest. Here is a brief outline of the contents.

In Chap. 1, we briefly describe pretest, shrinkage and penalty estimation strategies for estimating regression parameters in a multiple regression model.

In Chap. 2, linear shrinkage, and shrinkage pretest strategies for estimating the normal and Poisson means from a single sample have been presented. Large sample properties of the estimators are appraised and compared with the classical estimators.

Combining several sources of information can lead to improved estimates of the parameters of interest. In Chap. 3, efficient estimation strategies based on pretest and James-Stein principles for pooling data from several sources have been discussed. Simultaneous estimation of several coefficients of variation has been considered to illustrate the usefulness of pretest and shrinkage estimates.

In Chap. 4, shrinkage and pretest estimation in a multiple regression model have been discussed. Several penalty estimators such as LASSO, adaptive LASSO, and SCAD estimators have been presented. Monte Carlo simulation studies are provided and applications of suggested estimators using real data examples are given.

In Chap. 5, shrinkage, pretest, and penalty estimation strategies have been extended to partially linear regression models. The properties of the risk of the estimators have been studied both theoretically and through Monte Carlo studies.

In Chap. 6, shrinkage and penalty estimation strategies have been studied in a Poisson regression model. The performances of the estimators have been appraised through real data example as well as Monte Carlo simulation experiments.

Acknowledgments

I would like to express my appreciation to my colleagues, staff members, and several Ph.D. students for their interest and support in preparing this book. More specifically, I am grateful to Dr. K. Doksum, my former colleague, Drs. M. Hlynka, and A. Hissien for proofreading the earlier draft. I would also like to express my thanks to my former Ph.D. students, Drs. S. Chitsaz, S. Fallahpour, S. Hossain for their valuable assistance in the preparation of the manuscript, M with special thanks to Dr. E. Raheem, who made significant contributions in compiling this book. This book would not have been possible without the assistance of the incredible Springer team: E. Best, Dr. E. Hiripi, and Dr. V. Rosteck.

Finally, my heartfelt thanks you to my wife Ghazala, my son Jazib, and my daughter Feryaal. You are a constant source of joy and a great source of encouragement and motivation. I could not have done it without you!

Canada, July 2013 S. Ejaz Ahmed

Contents

Chapter 1
Estimation Strategies

Abstract Inference under uncertain prior information has been common for several decades in statistical and related literature. In the context of a regression model, we introduce the basic notion of full model, submodel, pretest, and shrinkage estimation strategies. We briefly discuss some penalty estimators and compare it with nonpenalty estimators.

Keywords Regression model · Pretest and shrinkage estimation · Penalty estimation

1.1 Introduction

For the past few decades, the use of *uncertain prior information* (UPI) or nonsample information (NSI) has been common in the statistical inference of *conditionally specified models*. Uncertain prior information is generally incorporated in a model via a conjectured constraint on the model parameters, thus giving rise to candidate submodels. The constraint on the parameters is usually obtained through expert opinion or from past data. Alternatively, candidate submodels can be obtained by using existing variable selection techniques, such as AIC or BIC, among others, when models are assumed to be sparse. Generally speaking, a submodel is more feasible and practical in a host of applications. When the conjectured constraint holds, or the submodel based on variable selection approach is true, analysis of such submodels leads to efficient statistical inferences than would be obtained through the full model. No matter how the submodel is obtained, either by variable selection technique or by imposing the constraints based on expert opinion, the estimators based on a submodel may become considerably biased and inefficient if the submodel is not the true representative of the data at hand. Recently, simultaneous variable selection and estimation of submodel parameters has become quite popular. However, such procedures may also produce biased estimators of the candidate submodel parameters. An example of a procedures along this line of thought is *penalty estimation* and their variants.

DOI: 10.1007/978-3-319-03149-1_1, © The Author(s) 2014

A different, but related breed of estimators, existed in the literature for more than five decades, is the so-called pretest and shrinkage estimators which attempt to incorporate uncertain prior information into the estimation procedure. Bancroft (1944) suggested the *pretest estimation* (PE) strategy. The pretest estimation approach uses a test to decide the estimator based on either the full model or a submodel. Broadly speaking, Bancroft (1944) suggested two seemingly unrelated problems on pretesting strategy. One is a data pooling problem from various sources based on a pretest approach and other one is related to simultaneous model selection and pretest estimation problem in regression model. Since then a bulk of research has been done using the pretest estimation procedure in many applications. However, the pretest estimator can also be seen as arising indirectly from the study of the admissibility of the sample mean vector in a multivariate normal distribution (Stein 1956; James and Stein 1961). In fact, the James–Stein shrinkage estimation technique may be regarded as a smooth version of the Bancroft (1944) pretest estimation procedure.

Since its inception, both pretest and shrinkage estimation strategy has received considerable attention from researchers and practitioners. More importantly, it has been analytically demonstrated that shrinkage estimation strategy outshines the classical estimators in many scenarios. During the past two and half decades, Ahmed and his co-researchers, among others, have demonstrated that shrinkage estimators dominate the classical estimators in terms of *mean squared error* (MSE) for a host of statistical models. A detailed description of shrinkage estimation and large sample estimation techniques in a regression model can be found in Ahmed (1997). Pretest, shrinkage and likelihood-based methods continue to play vital roles in statistical inference. These strategies can be used for both variable selection and post estimation. Further, pretest and shrinkage methods provide extremely useful techniques for combining data from various sources. For instance, estimation methodologies in the arena of meta analysis are essentially of this type.

To fix the idea of pretest and shrinkage estimation, let us consider for the rest of this chapter the estimation problem in a multiple linear regression model and the remaining discussion follows. If $\boldsymbol{Y} = (y_1, y_2, \ldots, y_n)'$ is a vector of responses, and the superscript ($'$) denotes the transpose of a vector or matrix, and $\boldsymbol{X}$ is an $n \times p$ fixed design matrix, $\boldsymbol{\beta} = (\beta_1, \ldots, \beta_p)'$ is an unknown vector of parameters, then we have the following regression model

$$\boldsymbol{Y} = \boldsymbol{X}\boldsymbol{\beta} + \boldsymbol{\varepsilon}, \tag{1.1}$$

where $\boldsymbol{\varepsilon} = (\varepsilon_1, \varepsilon_2, \ldots, \varepsilon_n)'$ is the vector of unobservable random errors.

Suppose that we have a set of covariates to fit a regression model for the purpose of predicting a response variable. If a *a priori* is known or suspected that a subset of the covariates do not significantly contribute to the overall prediction of the average response, they may be left aside and a model without these covariates may be considered (sparse model). In other situations, a subset of the covariates may be considered as a nuisance, meaning that they are not of primary interest to the researcher but they cannot be completely ignored either (model with nuisance parameters). In such cases, the effect of the nuisance parameters must be taken into account in estimating the coefficients of the remaining regression parameters. A candidate model for the data that involves only the important covariates (or active predictors) in predicting the response is called a restricted model or a submodel since we use a subset of the covariates after putting some restrictions on the nuisance covariates. On the other hand, a model that takes all the variables into consideration is termed as the unrestricted model/candidate full model or simply the full model.

Full Model Estimation

Based on the sample information only, the full model or unrestricted estimator (UE) of the regression coefficient is given by

$$\hat{\boldsymbol{\beta}}^{\text{UE}} = (\boldsymbol{X}'\boldsymbol{X})^{-1}\boldsymbol{X}'\boldsymbol{Y}.$$

Uncertain Prior Information

Suppose based on prior information it is possible to restrict the regression coefficients to a given subspace as follows:

$$\boldsymbol{H}\boldsymbol{\beta} = \boldsymbol{h},$$

where, $\boldsymbol{H}$ is a known matrix and $\boldsymbol{h}$ is a vector of known constants.

Auxiliary Information

A more interesting application of the above restriction is that $\boldsymbol{\beta}$ can be partitioned as $\boldsymbol{\beta} = (\boldsymbol{\beta}_1', \boldsymbol{\beta}_2')'$. The subvectors $\boldsymbol{\beta}_1$ and $\boldsymbol{\beta}_2$ are assumed to have dimensions p_1 and p_2 respectively, with $p_1 + p_2 = p$. In high-dimensional data analysis, it is assumed that the model is sparse. In other words, it is plausible that $\boldsymbol{\beta}_2$ may be set to a null vector. This auxiliary information regarding subvector $\boldsymbol{\beta}_2$ can be obtained by applying existing variable selection procedures.

Submodel Estimation

Thus, under the subspace restriction the submodel estimator or restricted estimator (RE) is given by

$$\hat{\boldsymbol{\beta}}^{\text{RE}} = \hat{\boldsymbol{\beta}}^{\text{UE}} - (\boldsymbol{X}'\boldsymbol{X})^{-1}\boldsymbol{H}'(\boldsymbol{H}(\boldsymbol{X}'\boldsymbol{X})^{-1}\boldsymbol{H}')^{-1}(\boldsymbol{H}\hat{\boldsymbol{\beta}}^{\text{UE}} - \boldsymbol{h}).$$

Similarly, a submodel estimator can be constructed using auxiliary information, that is $\boldsymbol{\beta}_2 = \boldsymbol{0}$.

1.1.1 Pretest and Shrinkage Estimation

In a regression setup, pretest and shrinkage estimation arise when we have prior information about a subset of the available covariates. The prior information may or may not positively contribute to the estimation process. Nevertheless, it may be advantageous to use the UPI when sample information is limited or the quality of the data at hand is poor, or even the estimate based on sample data may not be reliable. It is however, important to note that the consequences of incorporating UPI depend on the quality or usefulness of the information being added into the estimation process. Any uncertain prior information may be tested before incorporating it into the model. Based on the idea of Bancroft (1944), uncertain prior information may be validated through a pretest, and depending on such validation, the information may be incorporated into the model as a parametric restriction, thus choosing between the submodel and the full model estimation procedure.

Once we have a submodel estimator along with the full model estimator, we can test the validity of the subspace information, using a suitable test statistic ϕ_n. In pretest estimation framework, we consider testing the restriction in the form of the following null hypothesis:

$$H_0 : \boldsymbol{H\beta} = \boldsymbol{h}.$$

The preliminary test estimator (PE) or simply pretest estimator for the regression parameter $\boldsymbol{\beta}$ is obtained as

$$\hat{\boldsymbol{\beta}}^{\mathrm{PE}} = \hat{\boldsymbol{\beta}}^{\mathrm{UE}} - (\hat{\boldsymbol{\beta}}^{\mathrm{UE}} - \hat{\boldsymbol{\beta}}^{\mathrm{RE}}) I(\phi_n < c_{n,\,\alpha}), \tag{1.2}$$

where $I(\cdot)$ is an indicator function, and $c_{n,\,\alpha}$ is the $100(1-\alpha)$ percentage point of the test statistic ϕ_n.

In the framework proposed by Stein (1956), the shrinkage estimator or Stein-type estimator takes a hybrid approach by shrinking the full model estimator to a plausible alternative estimator (submodel estimator). In this framework, the estimates are essentially being shrunken toward the submodel estimators.

A Stein-type shrinkage estimator (SE) $\hat{\boldsymbol{\beta}}^{\mathrm{S}}$ of $\boldsymbol{\beta}$ can be defined as

$$\hat{\boldsymbol{\beta}}^{\mathrm{S}} = \hat{\boldsymbol{\beta}}^{\mathrm{RE}} + (\hat{\boldsymbol{\beta}}^{\mathrm{UE}} - \hat{\boldsymbol{\beta}}^{\mathrm{RE}}) \left\{1 - k\phi_n^{-1}\right\}, \quad k \geq 3.$$

The *positive-part shrinkage estimator* (PSE) has been suggested in the literature. A PSE has the form

$$\hat{\boldsymbol{\beta}}_1^{\mathrm{S+}} = \hat{\boldsymbol{\beta}}_1^{\mathrm{RE}} + (\hat{\boldsymbol{\beta}}_1^{\mathrm{UE}} - \hat{\boldsymbol{\beta}}_1^{\mathrm{RE}}) \left\{1 - k\phi_n^{-1}\right\}^{+},$$

where we define the notation $z^+ = \max(0, z)$. This adjustment controls for the over-shrinking problem in SE.

Shrinkage estimation method combines estimates from the candidate full model and a submodel in an optimal way dominating the full model estimator.

For a sparse model, a schematic flowchart of shrinkage estimation is shown in Fig. 1.1.

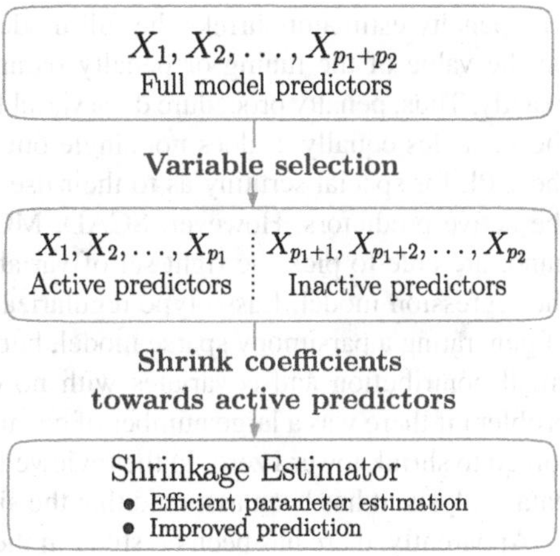

Fig. 1.1 Flowchart of shrinkage estimation strategy

1.1.2 Penalty Estimators

The penalty estimators are members of the penalized least squares family and they are obtained by optimizing a quadratic function subject to a penalty. Popular penalty estimators includes the least absolute and shrinkage operator (LASSO), adaptive LASSO, group LASSO, the smoothly clipped absolute deviation (SCAD), and minimax concave penalty (MCP), among others. They are so called because the penalty term is used in the penalized least squares to obtain the estimate of the regression parameters. The main idea in such estimation rule is that the regression coefficients with weak or no effects are shrunken toward the zero vector when model is sparse. The process often eliminates some of the covariates completely by shrinking their coefficients to exactly zero. Thus, the procedure performs variable selection and shrinkage together. In a sense, the procedure selects a submodel and estimates the regression parameters in the submodel. This technique is fruitful when the model is sparse, and applicable when number of predictors (p) is greater than the number of observations (n), a so-called ultrahigh dimensional model.

In shrinkage estimation strategies, if prior information about a subset of the covariates is available, then the estimates are obtained by incorporating the available information. However, in the absence of prior information, one might go with usual variable selection process to sort out the significant covariates. A typical variable selection process is likely to drop some of the covariates from the model. This is equivalent to having some prior knowledge about the dropped covariates that their coefficients are zero. Any suitable model selection criteria such as AIC, BIC, or penalty can be considered to achieve this objective.

Examining side-by-side the shrinkage and penalty estimation strategies, we see that the output of penalty estimation resembles shrinkage methods as it shrinks and selects the variables simultaneously. However, there is an important difference in how the shrinking works in penalty estimation when compared to the shrinkage estimation.

The penalty estimator shrinks the full model estimator toward zero and depending on the value of the tuning or penalty parameter, it sets some coefficients to zero exactly. Thus, penalty procedure does variable selection automatically by treating all the variables equally. It does not single out nuisance covariates, or for that matter, the UPI, for special scrutiny as to their usefulness in estimating the coefficients of the active predictors. However, SCAD, MCP, and adaptive LASSO, on the other hand, are able to pick the right set of variables while shrinking the coefficients of the regression model. Lasso-type regularization approaches have some advantages of generating a parsimony sparse model, but are not able to separate covariates with small contribution and covariates with no contributions. This could be a serious problem if there was a large number of covariates with small contributions and were forced to shrink toward zero. In the reviewed published studies on high-dimensional data analysis, it has been assumed that the signals and noises are well separated.

Apparently, there has been no study in the reviewed literature at one place which compares the relative performance of pretest, shrinkage, and penalty estimators. One of the objectives of the book is to present the relative properties of pretest, shrinkage, and penalty estimators in some useful models. Therefore, it is worth exploring the performance of the penalty and shrinkage estimators when it is suspected *a priori* that the parameters may be reduced to a subspace.

This presentation is fundamental, because pretest and Stein's methods may be appropriate for model selection problems and there is a growing interest in this area to fill the gap between two competitive strategies. Further, there is a dire need to investigate a more realistic case where there are small signals in the model and it cannot be effectively removed from the noise. The goal of this monograph is to discuss some of the issues involved in the estimation of the parameters in a host of models that may be over-parameterized by including too many variables in the model using shrinkage and pretest estimation strategies. For example, in genomics research, it is common practice to test a candidate subset of genetic markers for association with disease (e.g., Zeggini et al. 2007). Here the candidate subset is found in a certain population by doing genome wide association studies. The candidate subset is then tested for disease association in a new population. In this new population, it is possible that genetic markers not found in the first population are associated with disease. Shrinkage strategy is generally used to trade-off between bias and efficiency in a data adaptive way in order to provide meaningful analytical solutions to problems in genetic epidemiology and other applications.

1.2 Organization of the Book

This book is divided into six chapters.

In Chap. 2, we presented linear shrinkage and shrinkage pretest strategies for estimating the normal and Poisson means from a single sample, respectively. Asymptotic properties of the estimators have been proposed and compared with the classical estimators.

In Chap. 3, we extend the idea of pretest shrinkage strategies to a multiple sample case. Several estimators are proposed when multiple samples are available. For the multisample case, we demonstrate that the suggested shrinkage strategy is superior to the classical estimation strategy based on the sample data alone.

Shrinkage and pretest estimation in a multiple regression model are introduced in Chap. 4. Asymptotic bias and risk expressions of the estimators have been derived and the performance of shrinkage estimators is compared with the classical estimators using a real data example and through Monte Carlo simulation experiments. Several penalty estimators such as LASSO, adaptive LASSO, and SCAD estimators have been discussed. Monte Carlo simulation studies have been used to compare the performance of shrinkage and penalty estimators.

We discuss shrinkage, pretest, and penalty estimation, in partially linear model in Chap. 5. The risk properties of the estimators have been studied using asymptotic distributional risk and Monte Carlo simulation studies.

In Chap. 6 we considered shrinkage estimation in a Poisson regression model. This model assumes the response variable to have a Poisson distribution, and also that the logarithm of its expected value can be modeled by a linear combination of unknown parameters. We appraise the performance of shrinkage, penalty, and maximum likelihood estimators in this context with real data example and through Monte Carlo simulation experiments.

References

Ahmed, S. E. (1997). Asymptotic shrinkage estimation: The regression case. *Applied Statistical Science, II*, 113–139.

Bancroft, T. A. (1944). On biases in estimation due to the use of preliminary tests of significances. *Annals of Mathematical Statistics, 15*, 190–204.

James, W., & Stein, C. (1961). Estimation with quadratic loss. *Proceedings of the Fourth Berkeley Symposium on Mathematical Statistics and Probability* (pp. 361–379).

Stein, C. (1956). The admissibility of hotelling's T^2-test. *Mathematical Statistics, 27*, 616–623.

Zeggini, E., Weedon, M. N., Lindgren, C. M., Frayling, T. M., Elliott, K. S., Lango, H., et al. (2007). Replication of genome-wide association signals in uk samples reveals risk loci for type 2 diabetes. *Science, 316*(5829), 1336–1341.

Chapter 2
Improved Estimation Strategies in Normal and Poisson Models

Abstract In this chapter, we consider the estimation of the mean parameter from two commonly used statistical models in practice. In a classical approach, we estimate the parameter based on the available sample information at hand only. On the other hand, in Bayesian framework, we assume prior distribution on the parameter of interest to obtain an improved estimation. In semi-classical approach, we assume that an initial value of the parameter is available from past investigation or any other sources whatsoever. The main focus in this chapter is to combine sample information and nonsample information to obtain an improved estimator of the mean parameter of normal and Poisson models, respectively. To improve the estimation accuracy linear shrinkage and pretest estimation strategies are suggested. The performance of the suggested pretest estimator is appraised by using the mean squared error criterion. The relative efficiency of the suggested estimators with respect to a classical estimator is investigated both analytically and numerically. Not surprisingly, the linear shrinkage estimator outperforms its competitors when the nonsample information is nearly correctly specified. The pretest estimator is relatively more efficient than the classical estimator in the most interesting part of the parameter space. The suggested shrinkage pretest estimation strategy is easy to implement and does not require any tuning or hyperparameter. We strongly recommend using the shrinkage pretest estimation method for practical problems, since it does not drastically suffer postestimation bias or any other implications, unlike other methods, which fail to report the magnitude of the bias whether negligible or not negligible. The shrinkage pretest strategy precisely reports its strength and weakness.

Keywords Pretest and linear shrinkage estimation · Normal and poisson models · Bias and MSE

S. E. Ahmed, *Penalty, Shrinkage and Pretest Strategies*, SpringerBriefs in Statistics,
DOI: 10.1007/978-3-319-03149-1_2,

2.1 Introduction

In this chapter, we consider the estimation of the single-valued parameter θ based on a random sample of size n. To fix the idea, suppose we obtained observations $x_1, \ldots, x_n$ from a random variable X with probability density function (pdf) f. Further, let F be the cumulative distribution function (cdf) of the random variable X. Suppose θ is the mean parameter of $X \sim F$. The usual statistical problem concerns inference on the mean parameter θ. Although the estimation of the mean parameter is a classic problem, it leads to inference strategies for more involved and complex situations when θ may be a vector or a matrix of unknown quantities. We will address these situations later. Here, we confine ourselves to point estimation of the mean parameter θ. The statistical objective is to understand the unknown quantity θ in a model class $F = f_X\{x; \theta : \theta \in \Theta\}$ where Θ is the parameter space, has a finite dimension, and does not increase with the sample size.

The sample mean $\hat{\theta} = \overline{X} = \sum_{i=1}^{n} x_i/n$ is a traditional and natural point estimator of θ, and has been used in its own right before the beginning of statistics as a subject. This estimator is intuitively appealing and is a classic example of the *method of moments* estimation. Loosely speaking, we say that an estimator will be good if its values are near the true value of θ. Mathematically speaking, we designate an estimator as an unbiased estimator if $E(\hat{\theta}) = \theta$. We note the sample mean is an unbiased estimator of the mean parameter θ. Further, $\overline{X}$ is a *consistent* estimator of θ. It is very common to check the accuracy of the estimator using the notion of *mean squared error* (MSE), which is defined as

$$MSE(\hat{\theta}; \theta) = E\{(\hat{\theta} - \theta)^2\}.$$

The MSE can be conveniently decomposed into two components:

$$MSE(\hat{\theta}; \theta) = Var(\hat{\theta}) + \{bias(\hat{\theta}; \theta)\}^2,$$

where

$$Var(\hat{\theta}) = E[\{(\hat{\theta} - E(\hat{\theta})\}^2]$$

is a measure of the dispersion of $\hat{\theta}$ and

$$bias(\hat{\theta}; \theta) = E(\hat{\theta}) - \theta.$$

The first and foremost goal of statistical inference is to have an estimator with an MSE as small as possible. MSE can be used to measure the relative performance of two estimators for evaluating the same parameter. We will pick an estimator with the lowest MSE for further investigation. We consider the estimator $\hat{\theta}_1$ to be relatively better (*more efficient*) than $\hat{\theta}_2$ if

$$MSE(\hat{\theta}_1; \theta) \leq MSE(\hat{\theta}_2; \theta)$$

where strict inequality holds for at least one value of θ.

If a random sample is taken from a normal (Gaussian) distribution with known variance, then $\hat{\theta}$ is also the maximum likelihood estimator (MLE) of θ. Then

$$MSE(\hat{\theta}^{MLE}) = MSE(\overline{X}) = \frac{\sigma^2}{n}.$$

Thus, for a fixed sample size n, the MSE of $\hat{\theta}^{MLE}$ assumes a constant value of $\frac{\sigma^2}{n}$. Now, a natural question arises that asks how we can improve the efficiency of $\hat{\theta}^{MLE}$ for a given n. In this pursuit, one should look into the parameter space Θ for any additional information. It may be possible that $\theta \in \omega \subset \Theta$. This information can be used in forming a new estimation strategy in the hope that it will perform better than $\hat{\theta}^{MLE}$. In many real-life problems, practitioners have some conjectures about the value of parameter θ based on experience or familiarity with the experiment or survey under investigation. Suppose, in a biological experiment, a researcher is focused on estimating the growth rate parameter of a certain bacterium after applying some catalyst when it is suspected *a priori* that the growth parameter may be approximated by a specified value. In a controlled experiment, the ambient condition will not contribute to varying growth rates. Therefore, the biologist may have a good reason to suspect that the approximate value is the true growth rate parameter for her experiment. It is widely accepted that, in applied science, an experiment is often performed with some prior knowledge of the outcome, or to confirm a hypothetical result, or to re-establish existing results. It is reasonable, then, to move the point estimator of θ close to an approximation of θ, say θ_0. However, such information regarding the parameter is considered as *nonsample information* (NSI) or *uncertain prior information* (UPI).

2.2 Estimation Strategies

In this section, we provide some alternative estimation strategies to estimate the parameter on interest at hand, when some prior information about the parameter is readily available. We will evaluate the performance of the estimators using MSE criterion.

2.2.1 Linear Shrinkage Strategy

Our main focus is to improve the estimation of θ when it is generally assumed that the sample data may come from a distribution that is fairly close to a normal distribution. The data may be contaminated by a few observations, which will have a very negative impact on the sample estimate $\hat{\theta}$. Hence, in an effort to stabilize the parameter estimation of θ, we consider the problem of estimating θ when some prior information regarding the population parameter is available. In a number of real-world problems, the practitioner may have both an approximation of θ that provides a constant θ_o and sample information that provides a point estimator $\hat{\theta}$. The quality of θ_o is unknown; however, the analyst appreciates its ability to approximate θ. Our problem is to combine the approximation θ_o and the sample result $\hat{\theta}$. Consequently, we define a MSE-optimal linear shrinkage estimator as

$$\hat{\theta}^{LS} = c\theta_o + (1-c)\hat{\theta}, \tag{2.1}$$

where c denotes the shrinkage intensity. Ideally, the coefficient c is chosen to minimize the mean squared error. Further, c may also be defined as the degree of confidence in the prior information θ_o. The value of $c \in [0, 1]$ may be assigned by the experimenter according to confidence in the prior value of θ_o. If $c = 0$, then we use the sample data only.

The key advantage of this construction is that it outperforms the classical estimator $\hat{\theta}$ in some part of the parameter space. Estimators constructed as linear (or, more precisely, convex) combinations of other estimators or guessed values as in (2.1) are also known as *composite* estimators. However, the composite estimator $\hat{\theta}^{LS}$ can be interpreted as a *linear shrinkage estimator* (LSE), as it shrinks the sample estimator $\hat{\theta}$ toward θ_o. A lot of attention has been paid to this estimator, which is evident by the extensive publication in this area. Ahmed et al. (2011) developed asymptotic theory for this type of estimator. Ledoit and Wolf (2003) applied this strategy to estimate the covariance matrix. They suggested that shrinking the MLE of the covariance matrix toward structured covariance matrices can produce a relatively small estimation error in comparison with the MLE. Ahmed and Khan (1993), Ahmed (1991a, b, c), and others pointed out that such an estimator yields a smaller MSE when a priori information θ_o is correct or nearly correct. We will show that $\hat{\theta}^{S}$ will have a smaller MSE than $\hat{\theta}$ when θ is close to θ_o. However, $\hat{\theta}^{LS}$ becomes considerably biased and inefficient when the restriction may not be judiciously justified. Thus, the performance of this shrinkage procedure depends upon the correctness of the uncertain prior information.

To get some insight, let us consider the estimation of the normal mean parameter. Consider $X \sim N(\mu, \sigma^2)$ based on a sample size n, $\hat{\theta} = \hat{\theta}^{MLE} = \overline{X}$. For the sake of brevity, we assume that the population variance is known in the remaining discussion. The linear shrinkage estimator is given in (2.1). The $bias(\hat{\theta}^{LS}; \theta) = -c\delta$, where $\delta = \theta - \theta_o$. For $c \neq 0$, the bias is an unbounded function of δ. We know that the bias will not go away unless $\theta = \theta_o$. However, it is widely accepted or "generic" to incorporate some bias in the estimation process to achieve a decrease in MSE. This is referred to as the "bias-variance trade-off" in statistics and the related literature. To confirm this characteristic of such estimators, let us look at the MSE expression of $\hat{\theta}^{LS}$:

$$MSE(\hat{\theta}^{LS}; \theta) = \frac{\sigma^2}{n}\left[1 - c(2-c) + c^2\Delta\right],$$

where $\Delta = \frac{n\delta^2}{\sigma^2}$. Hence, we achieve our objective in lowering the MSE by incorporating the prior information in the estimation strategy. Clearly, $\hat{\theta}^{LS}$ will outperform the classical estimator $\overline{X}$ when Δ is small. However, for large values of δ the opposite conclusion will hold. It is possible to choose an estimator of optimal c that minimizes the MSE. The key question in this type of estimator is how to select a value for the shrinkage parameter c such that it may be regarded as an optimal estimator. In some situations, it may suffice to fix the parameter c at some given value. Another choice is to choose the parameter c in a data-driven fashion by explicitly minimizing a suitable risk function. A common but also computationally intensive approach is to estimate the minimizing c by using cross-validation. On the other hand, from a Bayesian perspective, one can employ the empirical Bayes technique to infer c. In this case, c is treated as a hyperparameter and may be estimated from the data by optimizing the marginal likelihood. Here, we treat c as the degree of trust in the prior information about the parameter $\theta = \theta_0$. In conclusion, it is clear that $\hat{\theta}^{LS}$ is a convex combination of $\hat{\theta}$ and $\overline{X}$ through the fixed value of $c \in (0, 1)$, and $\hat{\theta}^{LS}$ has a smaller MSE than $\overline{X}$ for small values of δ at the expense of its performance in the rest of the parameter space induced by δ. Not only that, its MSE becomes unbounded as $\delta \rightarrow \infty$. If the prior information is *bad* in the sense that the error in guessing is large, the linear shrinkage estimator will be inferior to the classical estimator. Alternatively, if the information is *good*, i.e., the guessing error is small, the linear shrinkage estimator offers a substantial gain over the classical estimator.

The above insight leads a to pretest estimation when the uncertain information is rather suspicious, and it is useful to construct a compromised estimator by performing a pretest on the prior information. Therefore, we obtain the pretest estimator as convex combinations of classical and linear shrinkage estimators via a test statistic. Useful discussions on some of the implications of pretest in parametric theory are given in Bancroft (1944), Sclove et al. (1972), Efron and Morris (1972), Judge and Bock (1978), and Ahmed and Saleh (1990), among others. For some asymptotic results on the subject we refer to Ahmed and Khan (1993), Ahmed (1991a, b, c), Ali and Saleh (1991), and Ahmed et al. (2011), among others.

2.2.2 *Shrinkage Pretest Strategy*

When the prior information is not trustworthy, it may be desirable to formulate a *pretest estimator* (PE) denoted by $\hat{\theta}^P$ which incorporates a pretest on θ_o. Thus, we consider the *shrinkage pretest estimator* (SPE) defined by

$$\hat{\theta}^P = \hat{\theta} I(\mathscr{D}_n \geq c_\alpha) + \theta_o I(\mathscr{D}_n < c_\alpha), \tag{2.2}$$

where $I(A)$ is the indicator function of set A, $\mathscr{D}_n = \frac{n(\overline{x}-\theta_o)^2}{\sigma^2}$ is the test statistic for the null hypothesis H_o: $\theta = \theta_o$, and c_α is the upper $100\alpha\%$ point of the distribution of $\mathscr{D}_n$. We consider testing H_o: $\theta = \theta_o$ against H_a: $\theta \neq \theta_o$ (or $\theta < \theta_o$ or $\theta > \theta_o$).

Further, PE can be written in a more computationally attractive form as follows:

$$\hat{\theta}^P = \hat{\theta} - (\hat{\theta} - \theta_o)I(\mathcal{D}_n < c_\alpha). \tag{2.3}$$

Essentially, we have replaced a fixed constant c in (2.1) with a dichotomous random quantity $I(\mathcal{D}_n < c_\alpha)$ to obtain (2.3). In return, we will achieve an estimator with a bounded MSE in terms of δ. It is important to remark that $\hat{\theta}^P$ performs better than $\hat{\theta}$ in some parts of the parameter space. The use of $\hat{\theta}^P$ may, however, be limited due to the large size of the pretest. Ahmed (1992) proposed a shrinkage pretest estimation strategy replacing θ_o by $\hat{\theta}^{LS}$ in (2.2) as follows:

$$\hat{\theta}^{SP} = \hat{\theta}I(\mathcal{D}_n \geq c_\alpha) + [(1-c)\hat{\theta} + c\theta_o]I(\mathcal{D}_n < c_\alpha), \tag{2.4}$$

or, equivalently,

$$\hat{\theta}^{SP} = \hat{\theta} - c(\hat{\theta} - \theta_o)I(\mathcal{D}_n < c_\alpha). \tag{2.5}$$

Ahmed (1992) discovered that $\hat{\theta}^{SP}$ significantly improves upon $\hat{\theta}^P$ in the size of the test, and dominates $\hat{\theta}$ in a large portion of the parameter space. However, the estimators based on the pretest strategy are biased.

The expression for the bias of $\hat{\theta}^{SP}$ is obtained with the aid of the following lemma from Judge and Bock (1978).

Lemma 2.1 *Let $Z \sim \mathcal{N}(\mu, 1)$. Then we have the following:*

$$E\{ZI(0 < Z^2 < z)\} = \mu P(\chi^2_{\nu,\frac{\mu^2}{2}} < z)$$

where $\chi^2_{\nu,\frac{\mu^2}{2}}$ has a chi-square distribution with ν degrees of freedom and noncentrality parameter $\frac{\mu^2}{2}$.

The bias expression for the shrinkage pretest estimator is given as follows:

$$bias(\hat{\theta}^{SP}; \theta) = -c\delta H_3(\chi^2_{1,\alpha}; \Delta),$$

where $H_3(\chi^2_{1,\alpha}; \Delta)$ is the noncentral chi-square distribution function with noncentrality parameter Δ and 3 degrees of freedom.

Now, we outline the derivation of the bias expression for $\hat{\theta}^{SP}$,

$$\begin{aligned} bias(\hat{\theta}^{SP}; \theta) &= E(\hat{\theta}^{SP} - \theta) \\ &= E(\hat{\theta} - \theta - c(\hat{\theta} - \theta_0)I(\mathcal{D}_n < c_\alpha)) \\ &= E(\hat{\theta} - \theta) - cE[(\hat{\theta} - \theta_0)I(\mathcal{D}_n < c_\alpha)] \\ &= -c\frac{\sigma}{\sqrt{n}}E(ZI(\mathcal{D}_n < c_\alpha)) \quad \text{(using Lemma 2.1)} \\ &= -c\delta H_3(\chi^2_{1,\alpha}; \Delta). \end{aligned}$$

For $c = 1$, $bias(\hat{\theta}^P; \beta) = -\delta H_3(\chi^2_{1,\alpha}; \Delta)$. Both pretest estimators ($\hat{\theta}^P$, and $\hat{\theta}^{SP}$) are unbiased when $\Delta = 0$. Also, they are asymptotically unbiased when $\delta \to \infty$ since $\lim_{\delta\to\infty} \delta H_3(\chi^2_{1,\alpha}; \Delta) = 0$. The bias functions of both pretest estimators increase to their maxima as Δ increases, then decrease towards 0 as Δ further increases. Also, it is seen from the bias expression that as the value of c becomes larger, the variation in bias becomes greater. Finally, $bias(\hat{\theta}^{SP}; \theta) < bias(\hat{\theta}^P; \theta)$ for $c \in (0, 1)$. One may also view c as a *bias reduction factor* or *bias controlling factor* in the pretest estimation.

The expression of $MSE(\hat{\theta}^{SP}; \theta)$ is readily obtained with the use of the following lemma from Judge and Bock (1978).

Lemma 2.2 *Let* $Z \sim \mathcal{N}(\mu, 1)$. *Then*

$$E\{Z^2 I(0 < Z^2 < z)\} = P\left(\chi^2_{3,\frac{\mu^2}{2}} < z\right) + \mu^2 P\left(\chi^2_{5,\frac{\mu^2}{2}} < z\right).$$

Using Lemmas (2.1) and (2.2), we present expressions for the MSE of the shrinkage pretest estimator.

$$\begin{aligned} MSE(\hat{\theta}^{SP}; \theta) &= \frac{\sigma^2}{n}\big[1 - c(2-c)H_3(\chi^2_{1,\alpha}; \Delta) \\ &\quad + c\Delta\{2H_3(\chi^2_{1,\alpha}; \Delta) - (2-c)H_5(\chi^2_{1,\alpha}; \Delta)\}\big]. \end{aligned}$$

A sketch of the derivation is given below:

$$\begin{aligned} MSE(\hat{\theta}^{SP}; \theta) &= E(\hat{\theta}^{SP} - \theta)^2 \\ &= E(\hat{\theta} - \theta - c(\hat{\theta} - \theta_0)I(\mathcal{D}_n < c_\alpha))^2 \\ &= E(\hat{\theta} - \theta)^2 + c^2 E[(\hat{\theta} - \theta_0)^2 I(\mathcal{D}_n < c_\alpha)] \\ &\quad - 2cE[(\hat{\theta} - \theta)(\hat{\theta} - \theta_0)I(\mathcal{D}_n < c_\alpha)] \\ &= \frac{\sigma^2}{n} + c^2\frac{\sigma^2}{n}E(Z^2 I(\mathcal{D}_n < c_\alpha)) \\ &\quad - 2cE[(\hat{\theta} - \theta_0 + \theta_0 - \theta)(\hat{\theta} - \theta_0)I(\mathcal{D}_n < c_\alpha)]. \end{aligned}$$

Now, the third term equals

$$\begin{aligned} &- 2cE[(\hat{\theta} - \theta_0)^2 I(\mathcal{D}_n < c_\alpha)] + 2cE[(\theta - \theta_0)(\hat{\theta} - \theta_0)I(\mathcal{D}_n < c_\alpha)] \\ &= -2c\frac{\sigma^2}{n}E(Z^2 I(\mathcal{D}_n < c_\alpha)) + 2c\frac{\sigma^2}{n}E(ZI(\mathcal{D} < c_\alpha)). \end{aligned}$$

Therefore, we have

$$
\begin{aligned}
MSE(\hat{\theta}^{SP};\theta) &= \frac{\sigma^2}{n} + c^2\frac{\sigma^2}{n}H_3(\chi^2_{1,\alpha};\Delta) + c^2\Delta H_5(\chi^2_{1,\alpha};\Delta) \\
&\quad - 2c\frac{\sigma^2}{n}H_3(\chi^2_{1,\alpha};\Delta) - 2c\Delta\frac{\sigma^2}{n}H_5(\chi^2_{1,\alpha};\Delta) + 2c\Delta\frac{\sigma^2}{n}H_3(\chi^2_{1,\alpha};\Delta) \\
&= \frac{\sigma^2}{n}\{1 + c(c-2)H_3(\chi^2_{1,\alpha};\Delta) + (c^2 - 2c)\Delta H_5(\chi^2_{1,\alpha};\Delta) \\
&\quad + 2c\Delta H_3(\chi^2_{1,\alpha};\Delta)\} \\
&= \frac{\sigma^2}{n}[1 - c(2-c) + c\Delta\{2H_3(\chi^2_{1,\alpha};\Delta) - (2-c)H_5(\chi^2_{1,\alpha};\Delta)\}].
\end{aligned}
$$

Ideally, one would like to find the optimal value of c such that MSE of $\hat{\theta}^{SP}$ is minimized. Theoretically this is possible by minimizing the MSE of $\hat{\theta}^{SP}$ with respect to c. Thus, we get the MSE-optimal shrinkage pretest weight

$$
c_{opt} = \frac{H_3(\chi^2_{1,\alpha};\Delta) + \Delta H_5(\chi^2_{1,\alpha};\Delta) - \Delta H_3(\chi^2_{1,\alpha};\Delta)}{H_3(\chi^2_{1,\alpha};\Delta) + \Delta H_5(\chi^2_{1,\alpha};\Delta)}.
$$

Not surprisingly, the optimum value of c is a function of Δ and is unknown, but it can be estimated using the sample data. However, we treat c as a constant and the remaining discussion is as follows. For $c = 1$, we get the *MSE* of $\hat{\theta}^P$ as follows:

$$
\begin{aligned}
MSE(\hat{\theta}^{P};\theta) &= \frac{\sigma^2}{n}\big[1 - H_3(\chi^2_{1,\alpha};\Delta) \\
&\quad + \Delta\{2H_3(\chi^2_{1,\alpha};\Delta) - H_5(\chi^2_{1,\alpha};\Delta)\}\big].
\end{aligned}
$$

2.2.3 Relative Performance

We note that the MSE of $\hat{\theta}$ is a constant line, while the MSE of $\hat{\theta}^{LS}$ is a straight line in terms of Δ, which intersects the MSE of $\hat{\theta}$ at $\Delta = (2-c)/c$. If the prior information is correct, then the MSE of $\hat{\theta}^{LS}$ is less than the MSE of $\hat{\theta}$. In addition,

$$
MSE(\hat{\theta}^{LS};\theta) \le MSE(\hat{\theta};\theta) \text{ when } \Delta \in \left[0, \frac{2-c}{c}\right].
$$

Hence, for Δ in this interval, $\hat{\theta}^{LS}$ performs better than $\hat{\theta}$. Alternatively, when Δ deviates from the origin beyond $(2-c)/c$, the MSE of $\hat{\theta}^{LS}$ increases and becomes unbounded. Thus, departure from the restriction is fatal to $\hat{\theta}^{LS}$ but is of less concern to $\hat{\theta}$.

In an attempt to identify some important characteristics of $\hat{\theta}^{LS}$, first note that

$$H_5(\chi^2_{1,\alpha}; \Delta) \leq H_3(\chi^2_{1,\alpha}; \Delta) \leq H_3(\chi^2_{1,\alpha}; 0), \tag{2.6}$$

for $\alpha \in (0, 1)$ and $\Delta \geq 0$. The first two terms of (2.6) converge to 0 as Δ approaches infinity. Using these results, we compare the MSE performance of $\hat{\theta}^{SP}$ with $\hat{\theta}$.

$$\begin{aligned} &MSE(\hat{\theta}^{SP}) \geq MSE(\hat{\theta}) \\ &\quad \text{if } \Delta \geq (2-c)H_3(\chi^2_{1,\alpha}; \Delta)\{2H_3(\chi^2_{1,\alpha}; \Delta) - (2-c)H_5(\chi^2_{1,\alpha}; \Delta)\}^{-1}. \end{aligned} \tag{2.7}$$

Thus, $\hat{\theta}^{SP}$ dominates $\hat{\theta}$ whenever

$$\Delta < (2-c)H_3(\chi^2_{1,\alpha}; \Delta)\{2H_3(\chi^2_{1,\alpha}; \Delta) - (2-c)H_5(\chi^2_{1,\alpha}; \Delta)\}^{-1}. \tag{2.8}$$

It is evident from (2.7) that MSE of $\hat{\theta}^{SP}$ is less than the MSE of $\hat{\theta}$ if Δ is equal to or near 0. As the level of the significance approaches one, the MSE of $\hat{\theta}^{SP}$ tends to the MSE of $\hat{\theta}$. Also, when Δ increases and tends to infinity, the MSE of $\hat{\theta}^{SP}$ approaches the MSE of $\hat{\theta}$. Further, for larger values of Δ, the value of the shrinkage pretest MSE increases, reaches its maximum after crossing the MSE of the classical estimator, and then monotonically decreases and approaches the MSE of $\hat{\theta}$. Therefore, there are points in the parameter space where the shrinkage pretest estimator has larger MSE than the classical estimator and a sufficient condition for this result to occur is that (2.7) holds. Figure 2.1 shows that the MSE of shrinkage pretest estimator for $c = 0.1, 0.3, 0.5, 0.7, 0.9$ and the MSE of $\hat{\theta}^{LS}$ estimator for $c = 0.5$ at some selected values of α. It appears from the figure that the smaller the value of α, the greater the variation is in the maximum and minimum values of the MSE of the shrinkage pretest estimator. On the other hand, MSE of $\hat{\theta}^{LS}$ is a linear function of Δ and increases without a bound as Δ increases.

Now, we investigate the MSE performance of $\hat{\theta}^P$.

$$\begin{aligned} &MSE(\hat{\theta}^{P}) \geq MSE(\hat{\theta}) \\ &\quad \Delta \geq (H_3(\chi^2_{1,\alpha}; \Delta)\{2H_3(\chi^2_{1,\alpha}; \Delta) - H_5(\chi^2_{1,\alpha}; \Delta)\}^{-1}. \end{aligned} \tag{2.9}$$

Thus, the range of the parameter space in (2.7) is smaller than that in (2.9). Therefore, the MSE of $\hat{\theta}^{SP}$ will be less than that of the MSE of $\hat{\theta}$ in a larger parameter space than that of $\hat{\theta}^P$. Hence, the shrinkage pretest estimator provides a wider range than that of the traditional pretest estimator in which it dominates the classical estimator $\hat{\theta}$. This indicates the superiority of the shrinkage pretest estimator over the pretest estimator.

The MSE difference

$$\begin{aligned} MSE(\theta^P; \theta) - MSE(\hat{\theta}^{SP}; \theta) = {} & \Delta\frac{\sigma^2}{n}\{2(1-c)H_3(\chi^2_{1,\alpha}; \Delta) \\ & - (1-c)^2 H_5(\chi^2_{1,\alpha}; \Delta)\} \\ & - \frac{\sigma^2}{n}(1-c)^2 H_3(\chi^2_{1,\alpha}; \Delta), \end{aligned} \tag{2.10}$$

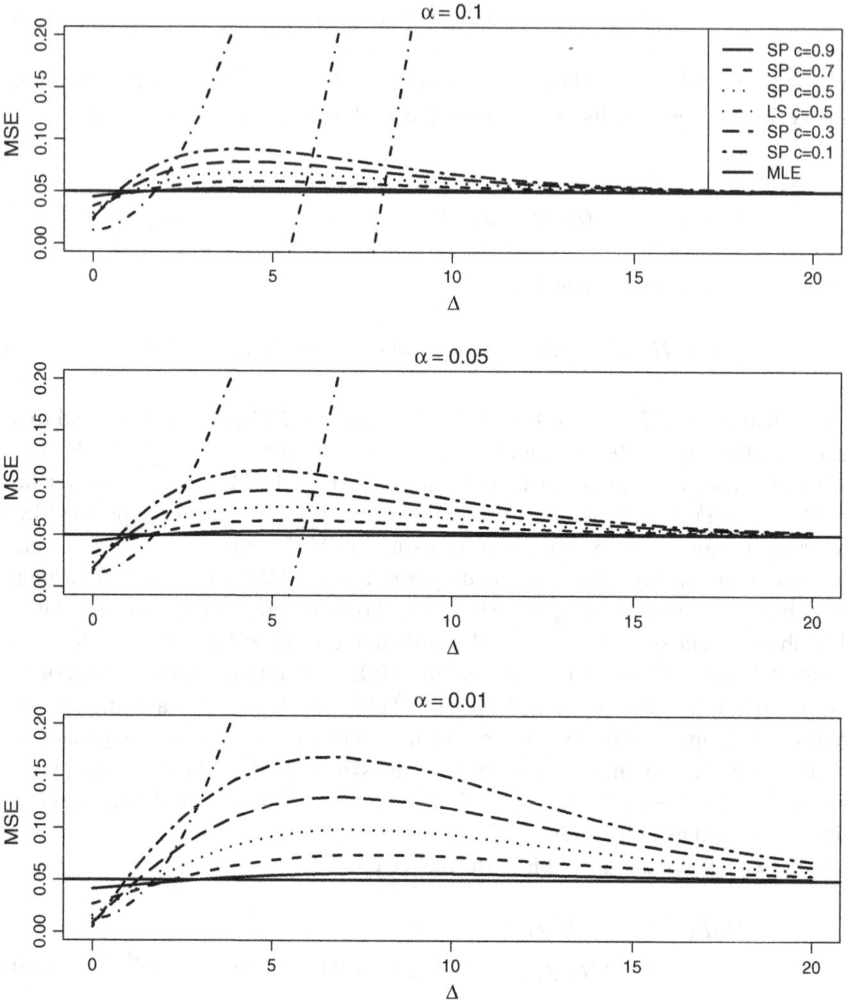

Fig. 2.1 Comparison of MSE for different values of Δ

suggests that $MSE(\hat{\theta}^P) \leq MSE(\hat{\theta}^{SP})$ whenever

$$\Delta \leq (1-c)H_3(\chi^2_{1,\alpha};\Delta)\{2H_3(\chi^2_{1,\alpha};\Delta)-(1-c)H_5(\chi^2_{1,\alpha};\Delta)\}^{-1}. \quad (2.11)$$

Thus, $\hat{\theta}^{SP}$ outshines $\hat{\theta}^P$ when

$$\Delta > (1-c)H_3(\chi^2_{1,\alpha};\Delta)\{2H_3(\chi^2_{1,\alpha};\Delta)-(1-c)H_5(\chi^2_{1,\alpha};\Delta)\}^{-1}. \quad (2.12)$$

It is seen that the MSE of $\hat{\theta}^P$ will be smaller than $\hat{\theta}^{SP}$ for small values of Δ, which may be negligible for larger values of c. Alternatively, when the value of Δ increases, then $\hat{\theta}^{SP}$ will dominate $\hat{\theta}^P$ in the rest of the parameter space. For a given c, let Δ_c be a point in the parameter space at which the MSE of $\hat{\theta}^{SP}$ and $\hat{\theta}^P$ intersect. Then, for $\Delta \in (0, \Delta_c]$, $\hat{\theta}^P$ performs better than $\hat{\theta}^{SP}$, while for $\Delta \in [\Delta_c, \infty)$, $\hat{\theta}^{SP}$ dominates $\hat{\theta}^P$. Further, for large values of c (close to 1), the interval $(0, \Delta_c]$ may not be significant. Nonetheless, $\hat{\theta}^P$ and $\hat{\theta}^{SP}$ share a common asymptotic property: as Δ grows and tends to infinity, their MSEs converge to a common limit, i.e., to the MSE of $\hat{\theta}$.

Hence, it is clear that none of the pretest estimators dominate each other, however, their MSE functions are bounded in Δ. Importantly, the shrinkage pretest estimator renders a wider range of values of the noncentrality parameter than the usual pretest estimator in which it dominates the classical estimator. However, it is important to remember that at $\Delta = 0$, the linear shrinkage estimator will be the best choice. Also, both pretest estimators have smaller MSE than that of $\hat{\theta}$ for small values of Δ.

2.2.4 *Size of the Pretest*

Estimation strategy based on pretesting is a function of the size of the pretest. For this purpose, we use the notion of relative MSE. The *relative MSE* (RMSE) of $\hat{\theta}$ to the estimator $\hat{\theta}^\star$ is defined by

$$RMSE(\hat{\theta}^\star : \hat{\theta}) = \frac{MSE(\hat{\theta}, \theta)}{MSE(\hat{\theta}^\star, \theta)}.$$

Keep in mind that a value of *RMSE* greater than 1 signifies improvement of $\hat{\theta}^\star$ over $\hat{\theta}$. The RMSE of $\hat{\theta}^{SP}$ to $\hat{\theta}$ is given by

$$RMSE(\hat{\theta}^{SP} : \hat{\theta}) = \frac{MSE(\hat{\theta}, \theta)}{MSE(\hat{\theta}^{SP}, \theta)} = \frac{1}{1 + g(\alpha, \Delta, c)}, \tag{2.13}$$

where

$$g(\alpha, \Delta, c) = \Delta c\left\{2H_3(\chi_1^2; \Delta) - 2(2-c)H_5(\chi_1^2, \Delta)\right\} - 2(2-c)H_3(\chi_1^2; \Delta). \tag{2.14}$$

Note that RMSE is a function of α, Δ, and c. This function, for $\alpha \neq 0$, has its maximum at $\Delta = 0$ with value

$$E_{\max} = \{1 - 2(2-c)H_3(\chi_1^2; 0)\}^{-1} (> 1).$$

Moreover, for fixed values of α and c, RMSE decreases as Δ increases from 0, crosses the line RMSE $= 1$, attains a minimum value at a point Δ_{min}, and then increases asymptotically to 1. However, for fixed c, E_{max} is a decreasing function of α while the minimum efficiency is an increasing function of α. On the other hand, for any fixed α, the maximum value of RMSE is a decreasing function of c and the minimum efficiency is an increasing function of c. The shrinkage factor c may also be viewed as a variation controlling factor among the maximum and minimum RMSE.

One method to determine α and c is to use a *maxmin* rule. For this, we allocate a value of the *minimum RMSE* (E_{min}) that we are willing to accept. Consider the set

$$A = \{\alpha, c | RMSE(\alpha^*, c^*, \Delta) \geq E_{min}, \ \forall \Delta\}.$$

The estimator is chosen, which maximizes $RMSE(\alpha, c, \Delta)$ over all $\alpha, c \in A$, and Δ. Thus, we solve for α^* and c^* such that

$$\sup_{\alpha, c \in A} \left\{ \inf_{\Delta} RMSE(\alpha^*, c^*, \Delta) \right\} = E_{min}. \tag{2.15}$$

For given $c = c_0$ we determine the value of α such that

$$\sup_{\alpha, c \in A} \left\{ \inf_{\Delta} RMSE\left(\alpha^*, c_0, \Delta\right) \right\} = E_{min}. \tag{2.16}$$

Table 2.1 provides the values of maximum RMSE (E_{max}), minimum RMSE (E_{min}), and the value of α^* for $c = 0.05, \ 0.1(0.1)1.0$.

Table 2.1 shows that, when c increases, the minimum relative MSE increases and maximum MSE decreases. Hence, c^* cannot satisfy (2.16) as it does not exist. The value of c can be determined by the researcher according to a prior belief about the uncertain prior information. However, we recommend the following two cases for selecting the size of the pretest:

Case 1: Suppose the experimenter does not know the size of the test but knows $c = c_o$ and wishes to accept an estimator which has relative MSE no less than E_{min}. Then the *maxmin* principle determines $\alpha = \alpha^*$ such that

$$RMSE(\alpha^*, c_o, \Delta) = E_{min}.$$

As an example, for $c = 0.5$ to achieve an RMSE of at least 0.72, Table 2.1 shows $\alpha = \alpha^* = 0.10$. Such a choice of α^* would yield an estimator with a maximum RMSE of 1.72 at $\Delta = 0$ and with a minimum guaranteed RMSE of 0.73. On the other hand, if the practitioner wishes to rely on data completely, then from Table 2.1 the size of the pretest will be approximately 0.25. Also, the maximum RMSE drops from 1.72 to 1.38. Thus, the use of a traditional pretest estimator is limited by the size of α, and the level of significance as compared to a shrinkage pretest estimator. Hence,

Table 2.1 Maximum and minimum RMSE of $\hat{\theta}^{SP}$ relative to $\hat{\theta}$

	c											
α	0.05	0.1	0.2	0.3	0.4	0.5	0.6	0.7	0.8	0.9	1.0	RMSE
0.01	1.09	1.21	1.49	1.87	2.41	3.19	4.32	5.99	8.25	10.67	11.83	E_{max}
	0.94	0.88	0.77	0.66	0.58	0.50	0.43	0.38	0.33	0.29	0.26	E_{min}
	10.00	9.00	9.00	9.00	9.00	8.00	8.00	8.00	8.00	8.00	8.00	$\Delta_{\min}$
0.025	1.08	1.18	1.42	1.73	2.13	2.64	3.30	4.08	4.91	5.60	5.88	$E_{\max}$
	0.95	0.90	0.81	0.72	0.64	0.57	0.51	0.45	0.41	0.36	0.33	$E_{\min}$
	8.00	8.00	8.00	8.00	7.00	7.00	7.00	7.00	7.00	7.00	6.00	$\Delta_{\min}$
0.05	1.07	1.15	1.35	1.58	1.85	2.17	2.53	2.90	3.24	3.49	3.58	$E_{\max}$
	0.96	0.92	0.84	0.77	0.71	0.64	0.58	0.53	0.48	0.44	0.40	$E_{\min}$
	7.00	7.00	7.00	7.00	6.00	6.00	6.00	6.00	6.00	6.00	6.00	$\Delta_{\min}$
0.10	1.05	1.11	1.25	1.40	1.55	1.72	1.89	2.04	2.16	2.24	2.27	$E_{\max}$
	0.97	0.94	0.88	0.83	0.78	0.73	0.68	0.63	0.59	0.55	0.51	$E_{\min}$
	6.00	6.00	6.00	6.00	6.00	5.00	5.00	5.00	5.00	5.00	5.00	$\Delta_{\min}$
0.15	1.04	1.09	1.18	1.29	1.39	1.49	1.59	1.67	1.73	1.77	1.79	$E_{\max}$
	0.97	0.95	0.91	0.87	0.82	0.78	0.74	0.70	0.66	0.63	0.59	$E_{\min}$
	5.00	5.00	5.00	5.00	5.00	5.00	5.00	5.00	5.00	5.00	5.00	$\Delta_{\min}$
0.20	1.03	1.07	1.14	1.21	1.28	1.35	1.41	1.46	1.50	1.53	1.53	$E_{\max}$
	0.98	0.96	0.93	0.89	0.86	0.83	0.79	0.76	0.73	0.69	0.66	$E_{\min}$
	5.00	5.00	5.00	5.00	5.00	5.00	5.00	5.00	5.00	4.00	4.00	$\Delta_{\min}$
0.25	1.02	1.05	1.11	1.16	1.21	1.26	1.30	1.33	1.36	1.37	1.38	$E_{\max}$
	0.98	0.97	0.94	0.92	0.89	0.86	0.83	0.81	0.78	0.75	0.72	$E_{\min}$
	5.00	5.00	5.00	5.00	5.00	5.00	4.00	4.00	4.00	4.00	4.00	$\Delta_{\min}$
0.30	1.02	1.04	1.08	1.12	1.16	1.19	1.22	1.24	1.26	1.27	1.27	$E_{\max}$
	0.99	0.98	0.95	0.93	0.91	0.89	0.87	0.84	0.82	0.80	0.77	$E_{\min}$
	5.00	5.00	5.00	5.00	4.00	4.00	4.00	4.00	4.00	4.00	4.00	$\Delta_{\min}$

the shrinkage pretest estimator has a remarkable edge over the pretest estimator with respect to the size of the pretest.

In real-life situations, the population variance σ^2 is rarely known to the experimenter. In the above discussion, we assumed that σ^2 is known primarily to keep the mathematical treatment simple and straightforward, and to keep clear and concise ideas and concepts about the notion of the proposed estimation strategies. However, one can easily implement the linear shrinkage and shrinkage pretest estimation strategies for the case when σ^2 is unknown.

2.2.5 Estimation Strategies when Variance is Unknown

Now, we provide a brief outline for estimating the normal mean parameter when σ^2 is unknown via shrinkage linear and shrinkage pretest estimation strategies. There will be no effect on the construction and MSE derivation of the linear shrinkage estimator. However, the pretest estimation strategy will be affected since the strategy depends on testing the mean when σ^2 is unknown. In this case, the test statistic will no longer follow a normal (χ^2) distribution; instead, it will have a $t(F)$-distribution. This will slightly effect the derivation of the MSE for the pretest estimation.

To fix the idea, let us consider $X \sim N(\theta, \sigma^2)$. The statistical objective is to estimate the mean parameter θ when σ is unknown using linear shrinkage and pretest estimation strategies. The linear shrinkage estimator given in (2.1) remains unchanged. However, the shrinkage pretest estimators are defined as follows:

$$\hat{\theta}^{SP} = \hat{\theta} - c(\hat{\theta} - \theta_o)I(\mathscr{D}_n < c_\alpha), \tag{2.17}$$

where

$$\mathscr{D}_n = \frac{n(\overline{x} - \theta_o)^2}{s^2},$$

and s^2 is the usual estimator for sample variance σ^2. Further, $c_\alpha = F_{1,n-1(\alpha)}$ is the upper α-level critical value from a central F-distribution.

The MSE expression for this estimator can readily be obtained by using Lemmas 2.1 and 2.2 as follows:

$$\begin{aligned} MSE(\hat{\theta}^{SP}; \theta) = \frac{\sigma^2}{n}\Big[1 - c(2-c)H_{3,\nu}\left(\frac{1}{3}F_{1,\nu(\alpha)}; \Delta\right) \\ + c\Delta\Big\{2H_{3,\nu}\left(\frac{1}{3}F_{1,\nu(\alpha)}; \Delta\right) - (2-c)H_{5,\nu}\left(\frac{1}{5}F_{1,\nu(\alpha)}; \Delta\right)\Big\}\Big], \end{aligned}$$

where $H_{\nu_1,\nu_2}(\cdot; \Delta)$ is the cumulative distribution function of a noncentral F-distribution with (ν_1, ν_2) degrees of freedom and noncentrality parameter $\Delta = \frac{n(\mu-\mu_0)^2}{\sigma^2}$.

The behavior of $\hat{\theta}^{SP}$ essentially remains the same as in the case when the variance was known and, hence, is not further pursued. For the selection of α, we suggest using $\alpha = 0.05$ for $\hat{\theta}^{SP}$. For a more precise selection of α, tables for maximum and minimum RMSE can also be prepared.

2.3 Estimation in Non-normal Models

The linear shrinkage and shrinkage pretest strategies can be implemented for non-normal models by using available asymptotic results for the classical methods, and then establishing the large sample asymptotic theory for the shrinkage and pretest estimators.

For the sake of brevity and to emphasize the application point of view, we next consider the estimation of the Poisson model parameter λ. The suggested methodology can easily be implemented for other commonly used discrete models and non-normal models.

2.3.1 A Discrete Model: Poisson Distribution

Let us consider the estimation of the Poisson mean parameter: $X \sim P(\lambda)$. Based on a sample size n, the MLE of λ is $\hat{\lambda} = \hat{\lambda}^{MLE} = \overline{X}$. Our main focus here is to improve the estimation of λ when it is generally assumed that the sample data may come from a distribution that is fairly close to the parameter λ of the Poisson distribution. However, the data may be contaminated by a few observations, which will have a very negative impact on the sample estimate $\hat{\lambda}$. Hence, in an effort to stabilize the parameter estimation of λ, we consider the problem of estimating λ when some prior information regarding the model parameter λ is available. In a number of real-world problems, the practitioner may have both an approximation of λ that provides a constant λ_o and sample information that provides a point estimator $\hat{\lambda}$. The quality of λ_o is unknown; however, the analyst appreciates its ability to approximate λ. The classical problem is to combine the approximation λ_o and the sample result $\hat{\lambda}$. Consequently, we consider estimators based on shrinkage and pretest estimation.

Suppose the analyst wishes to report the point estimator defined by the linear combination

$$\hat{\lambda}^{LS} = c\lambda_o + (1-c)\hat{\lambda}, \tag{2.18}$$

in which we would choose, in ideal circumstances, the coefficient c so as to minimize the mean squared error. Further, c may also be defined as the degree of confidence in the prior information λ_o. The value of $c \in [0, 1]$ may be assigned by the experimenter according to confidence in the prior value of λ_o. If $c = 0$, then we use the sample data only. We may choose an estimator of optimal c that minimizes the MSE. However, the optimal value of c depends on the unknown parameter λ and thus it is not accessible. As we pointed out earlier, such an estimator yields a smaller mean squared error when a prior information λ_o is correct or nearly correct. We will demonstrate that $\hat{\lambda}^{LS}$ will have a smaller MSE than $\hat{\lambda}$ when λ is close to λ_o. However, $\hat{\lambda}^{LS}$ becomes considerably biased and inefficient when the restriction may not be judiciously justified. Thus, the performance of this shrinkage procedure depends upon the correctness of the uncertain prior information. As such, when the prior information is not trustworthy, it may be desirable to formulate a shrinkage pretest estimator denoted by $\hat{\lambda}^{SP}$ which incorporates a pretest on λ_o. Thus, we consider the shrinkage pretest estimator which is defined by

$$\hat{\lambda}^{SP} = \hat{\lambda} I(\mathscr{L}_n \geq c_\alpha) + [(1-c)\hat{\lambda} + c\lambda_o] I(\mathscr{L}_n < c_\alpha), \tag{2.19}$$

where $I(A)$ is the indicator function of set A and $\mathscr{L}_n$ is the test statistic for the null hypothesis H_o: $\lambda = \lambda_o$,

$$\mathscr{L}_n = \frac{\{\sqrt{n}(\hat{\lambda} - \lambda_o)\}^2}{\hat{\lambda}}. \tag{2.20}$$

We consider testing H_o: $\lambda = \lambda_o$ against H_a: $\lambda \neq \lambda_o$. For moderate and large values of n and under the null hypothesis, the test statistic $\mathscr{L}_n$ follows a χ^2-distribution with one degree of freedom, which provides the asymptotic critical values. Thus, the critical value c_α of $\mathcal{L}_n$ may be approximated by $\chi^2_{1,\alpha}$, the upper $100\alpha\%$ critical value of the χ^2 distribution with 1 degree of freedom.

Note that the above result is based on the asymptotic normality of $\hat{\lambda}$. Consequently, the properties of the proposed improved estimators will be of an asymptotic nature.

2.3.2 Local Alternative and Asymptotic Criterion

It is important to note that for a fixed alternative that is different from the null hypothesis, the power of the test statistics will converge to one as $n \to \infty$. Hence, to explore the asymptotic power properties of $\mathcal{L}_n$ and to avoid the asymptotic degeneracy, we specify a sequence of local alternatives. Here, the local alternative setting is more reasonable since estimators based on the pretest principle are usually useful in the cases where λ and λ_0 are close. Therefore, for a given sample of size n a sequence $\{K_n\}$ of local alternatives is considered which is given by

$$K_n : \lambda_n = \lambda_o + \frac{\delta}{\sqrt{n}}. \tag{2.21}$$

Here δ is a fixed real number. Stochastic convergence of $\hat{\lambda}$ to the parameter λ ensures that $\hat{\lambda} \xrightarrow{p} \lambda$ under local alternatives as well, where the notation $\xrightarrow{p}$ means *convergence in probability*.

Furthermore, it is rather sensible to establish and compare the asymptotic properties of listed competitive estimators under local alternatives. We define an *asymptotic mean squared error* (AMSE) as the limit of the MSE for large n computed under the local alternatives. Similarly, the *asymptotic bias* (AB) of an estimator may be defined as the limit of the bias.

2.3.3 Asymptotic Bias and Asymptotic Mean Squared Error

The asymptotic bias of an estimator $\tilde{\theta}$ of the parameter θ is defined as

$$AB(\tilde{\theta}; \theta) = \lim_{n\to\infty} E\{\sqrt{n}(\tilde{\theta} - \theta)\}. \tag{2.22}$$

Under local alternatives, $AB(\hat{\lambda}^{LS}) = -c\,\delta$. Clearly, this is an unbounded function of δ. The expression of the asymptotic bias for the shrinkage pretest estimator is

$$AB(\hat{\lambda}^{SP}) = -c\,\delta G_3(\chi^2_{1,\alpha}; \Delta), \tag{2.23}$$

where $G_\nu(\cdot;\Delta)$ is the cumulative distribution function of a noncentral chi-square distribution with ν degrees of freedom and noncentrality parameter Δ. The expression of $AB(\hat{\lambda}^{SP})$ is obtained with the aid of the Lemma 2.1. Since $\lim_{\delta\to\infty} \delta G_3(\chi^2_{1,\alpha};\Delta) = 0$, we can safely conclude that $\hat{\lambda}^{SP}$ is asymptotically unbiased, with respect to δ. For $c = 1$, $AB(\hat{\lambda}^P) = -\delta G_3(\chi^2_{1,\alpha};\Delta)$. The quantities $AB(\hat{\lambda}^{SP})$ and $AB(\hat{\lambda}^P)$ are 0 at $\Delta = 0$. The bias functions of both pretest estimators increase to their maxima as Δ increases, then decrease toward 0 as Δ further increases. Also, it is seen from the AB expression that the larger the value of c, the greater the variation in the bias. Thus, the analysis remains the same as for a fixed sample size.

Under the local alternatives in (2.21), we present the expressions for the *AMSE* for the estimators under consideration.

$$AMSE(\hat{\lambda}^S;\lambda) = AMSE(\hat{\lambda};\lambda)[1 - c(2-c) + c^2\Delta], \tag{2.24}$$

where $AMSE(\hat{\lambda};\lambda) = \lambda$.

$$\begin{aligned} AMSE(\hat{\lambda}^{SP};\lambda) = AMSE(\hat{\lambda};\lambda)[1 &- c(2-c)G_3(\chi^2_{1,\alpha};\Delta) \\ &+ c^2\Delta\{2G_3(\chi^2_{1,\alpha};\Delta) - (2-c)G_5(\chi^2_{1,\alpha};\Delta)\}]. \end{aligned} \tag{2.25}$$

The expression of $AMSE(\hat{\lambda}^{SP})$ is readily obtained with the use of the Lemma 2.2.

The shrinkage pretest strategy preserves its MSE characteristics in the non-normal (large-sample) scenario. The relative performance of the shrinkage pretest estimator to the classical estimator is identical to the findings of Sect. 2.2.3.

For $c = 1$ we get the *AMSE* of $\hat{\lambda}^P$ as follows:

$$\begin{aligned} AMSE(\hat{\lambda}^P;\lambda) = \lambda[1 + \Delta\{2G_3(\chi^2_{1,\alpha};\Delta) - G_5(\chi^2_{1,\alpha};\Delta)\} \\ \Delta G_3(\chi^2_{1,\alpha};\Delta)] \end{aligned} \tag{2.26}$$

and $AMSE(\hat{\lambda}^P;\lambda) \geq AMSE(\hat{\lambda}^{SP};\lambda)$. Accordingly,

$$\Delta \geq G_3(\chi^2_{1,\alpha};\Delta)\{2G_3(\chi^2_{1,\alpha};\Delta) - G_5(\chi^2_{1,\alpha};\Delta)\}^{-1}. \tag{2.27}$$

Again, similar conclusions can be drawn regarding the competitive performance of two pretest estimators as of those in Sect. 2.2.3.

2.4 Chapter Summary

We presented improved estimation strategies for estimating normal and Poisson means by utilizing the nonsample information (using approximate values of the parameter), respectively. We suggested linear shrinkage and shrinkage pretest strategies for estimating the mean parameters. The properties of the proposed estimators were appraised and compared with the classical estimator using bias and MSE measures. The analysis revealed that the shrinkage pretest estimator is a bounded function of approximation error, and it offers substantial MSE reduction when the approximation is nearly correct. The suggested shrinkage pretest estimation strategy is easy to implement and does not require any tuning or hyperparameter. It also gives a comparable performance in simulation. We strongly recommend using the shrinkage pretest estimation method for practical problems, since it does not suffer drastically postestimation bias or any other implications, unlike other methods which fail to report the magnitude of the bias, whether negligible or not negligible. The shrinkage pretest strategy precisely reports its strengths and weaknesses.

In the next chapter, we extend the pretest shrinkage strategy to a multisample case. In this scenario, we also suggest a nonlinear shrinkage estimation which resembles the Stein-rule estimation. More importantly, we will demonstrate that the suggested shrinkage strategy is superior to the classical estimation strategy.

References

Ahmed, S. E. (1991a). Combining poisson means. *Communications in Statistics: Theory and Methods*, *20*, 771–789.

Ahmed, S. E. (1991b). A note on the estimation of proportion in binomial population. *Pakistan Journal of Statistics*, *6*, 63–70.

Ahmed, S. E. (1991c). Use of a priori information in the estimation of poisson parameter. *Soochow Journal of Mathematics*, *16*, 185–192.

Ahmed, S. E. (1992). Shrinkage preliminary test estimation in multivariate normal distributions. *Journal of Statistical Computation and Simulation*, *43*, 177–195.

Ahmed, S. E., & Khan, S. M. (1993). Improved estimation of the poisson parameter. *Statistica, anno LIII, 2*, 268–286.

Ahmed, S. E., Omar, M. H., & Joarder, A. H. (2011). Stabilizing the performance of kurtosis estimator of multivariate data. *Communications in Statistics—Simulation and Computation*, *40*, 3540–3554.

Ahmed, S. E., & Saleh, A. K. Md. E. (1990). Estimation strategies for the intercept vector in a simple linear multivariate normal regression model. *Computational Statistics and Data Analysis*, *10*, 193–206.

Ali, A. M., Saleh, A. K. Md. E. (1991). Preliminary test and empirical bayes approach to shrinkage estimation of regression parameters. *Journal of Japan Statistical Society*, *21*(1), 401–415.

Bancroft, T. A. (1944). On biases in estimation due to the use of preliminary tests of significances. *Annals of Mathematical Statistics*, *15*, 190–204.

Efron, B., & Morris, C. (1972). Empirical bayes on vector observations—an extension of stein's method. *Biometrika*, *59*, 335–347.

Judge, G. G., & Bock, M. E. (1978). *The statistical implications of pre-test and Stein-rule estimators in econometrics*. Amsterdam: North Holland.

Ledoit, O., & Wolf, M. (2003). Honey, I shrunk the sample covariance matrix. In *UPF Economics and Business Working Paper No. 691*.

Sclove, S. L., Morris, C., & Radhakrishnan, R. (1972). Non optimality of preliminary test estimation for the multinormal mean. *The Annals of Mathematical Statistics*, *43*, 1481–1490.

Chapter 3
Pooling Data: Making Sense or Folly

Abstract Pooling data from various sources to improve the parameter estimation is an important problem from the practitioner's perspective. If the pooling procedure is carried out judiciously, a much more efficient estimation strategy can be achieved for the targeted parameter. However, it is imperative that underlying assumptions for pooling the data are investigated thoroughly before merging the data into a single data set, and suggesting a pooled estimator based on a grand data. In this chapter, we explore various estimation strategies for pooling data from several sources. We suggest some efficient estimation strategies based on pretest and James–Stein principles. We consider simultaneous estimation of several coefficients of variation to demonstrate the power and beauty of pretest and shrinkage estimation in pooling data. We investigate the asymptotic and finite sample properties of these estimators using mean squared error criterion. We showcase that the shrinkage estimators based on the James–Stein rule dominate the benchmark estimator of coefficients of variation.

Keywords Coefficients of variation · Pretest and shrinkage estimation · Meta-analysis · Bias and efficiency

3.1 Introduction

In this chapter, we consider the problems that may occur when estimating the parameter vector from several models of interest based on differing sample sizes. We plan to investigate whether or not all samples are drawn from the same population, and this information is used to improve the usual estimates of the parameters of interest. The estimators are based on the shrinkage, pretest, and James–Stein rules. Instead of using the classical several normal means estimation problem, we consider simultaneous estimation of coefficients of variation from several independent normal models. The proposed estimation strategies can be effectively implemented to a host of estimation problems and various statistical models. In that sense, the suggested

S. E. Ahmed, *Penalty, Shrinkage and Pretest Strategies*, SpringerBriefs in Statistics,
DOI: 10.1007/978-3-319-03149-1_3,

estimation strategies are very general; however, their properties are developed in the context of estimating coefficients of variation.

The coefficient of variation (CV) of a random variable X with mean μ and standard deviation σ is defined by the ratio σ/μ with $\mu \neq 0$. For convenience of notation, CV will be denoted by the Greek letter γ. This ratio is a measure of relative dispersion, and is useful in many applications. Further, noting that the population coefficient of variation γ is a pure number and is free from units of any measure. The coefficient of variation can be used to compare the variability of two different populations. It is sometimes a more informative quantity than σ. For example, a value of 15 for σ has not much meaning unless it can be compared with μ. If σ is known to be 23 and μ is 9500, then the magnitude of the variation is small relative to its mean.The coefficient of variation play an important role in financial markets and statistical quality control, and other fields. As an example, in the study of the precision of a measuring instrument, engineers are typically more interested in estimating γ than estimating σ in its own right.

Now, we turn our attention to multi-sample estimation problem. For example the multiple independent random samples are obtained at different time points or from populations that have similar characteristics to estimate the CV. The experimenters are interested in the analysis of data sets collected in separate studies of the same phenomenon. Data sets analyzed in such a manner are so-called *meta-analysis*. However, meta-analysis does not go beyond the assumption of equality of the parameters at hand.

Generally speaking, the population coefficient of variation is fairly stable over time and over similar types of characteristics. In this data collection process, it is logical to consider that $\gamma_1 = \gamma_2 = \cdots = \gamma_k$. In the reviewed literature, the assumption of the homogeneity of coefficients of variation is common in biological and agricultural experiments. The main theme of this chapter is to consider the problem of simultaneous estimation of γ_i, $i = 1, 2, \ldots, k$ when the assumption of equality of the parameters may or may not hold. Thus, we showcase a much broader and unified estimation strategy for the estimation of several parameters.

To begin the work let us consider here that $Y_{i1}, Y_{i2}, \ldots, Y_{in_i}$ $(i = 1, 2, \ldots, k)$ is a random sample of size n_i taken from the ith population modeled by a normal distribution. Let us borrow matrix algebra notation and define the mean parameter vector $\boldsymbol{\mu} = (\mu_1, \mu_2, \ldots, \mu_k)'$ and covariance matrix $\sigma_i^2 \mathbf{I}_{(k\times k)}$, where $\mathbf{I}$ is an identity matrix. The individual CV is defined by $\gamma_i = \frac{\sigma_i}{\mu_i}$, $\mu_i \neq 0$, $i = 1, 2, \ldots, k$. In reality μ_i and σ_i^2 are usually unknown, the *unrestricted estimator* (UE) of μ_i and σ_i^2 are denoted by $\hat{\mu}_i$ and $\hat{\sigma}_i^2$, respectively, and

$$\hat{\mu}_i = \frac{1}{n_i}\sum_{j=1}^{n_i} Y_{ij} \quad \text{and} \quad \hat{\sigma}_i^2 = \frac{1}{n_i}\sum_{j=1}^{n_i}(Y_{ij} - \hat{\mu}_i)^2. \tag{3.1}$$

Similarly,

$$\hat{\boldsymbol{\gamma}} = (\hat{\gamma}_1, \ldots, \hat{\gamma}_k)',$$

where $\hat{\gamma}_i = \hat{\sigma}_i / \hat{\mu}_i, i = 1, 2, \ldots k$. Further,

$$\sqrt{n_i}\left(\frac{\hat{\gamma}_i^2}{2} + \hat{\gamma}_i^4\right)^{-\frac{1}{2}} (\hat{\gamma}_i - \gamma_i) \xrightarrow{D} \mathcal{N}(0, 1), \tag{3.2}$$

where $\xrightarrow{D}$ means convergence in distribution, see Ahmed (2002).

The plan for this chapter is as follows: in Sect. 3.2, we introduce various improved estimation strategies for CV parameter vector. We then present some useful asymptotic results which are given in Sect. 3.3. In Sect. 3.4, the expressions and analysis for asymptotic bias and risk of the estimators are provided. The result of a limited simulation study is presented in Sect. 3.5.

3.2 Efficient Estimation Strategies

We introduce improved estimation strategies for the parameter vector $\boldsymbol{\gamma} = (\gamma_1, \ldots, \gamma_k)'$ which incorporates both sample information and the conjecture so that all the coefficients of variation may be the same or similar. In a sense, we are describing two models here:

- A full model with $\gamma_i, i = 1, \ldots, k$ parameters of interest to estimate based on sample information only.
- A candidate submodel with one common parameter γ, that is $\gamma_1 = \gamma_2 = \cdots = \gamma_k = \gamma$ to be estimated based on sample size n.

Clearly, the estimator based on a submodel will outperform the estimators on a full model, if the homogeneity assumption of the parameter holds. However, an important statistical question is what happens if this assumption judiciously cannot be justified. Regardless, whether a submodel is selected by a human eye or by any other approach, the estimation consistency of selected submodel parameter(s) is questionable. We address this estimation problem in this chapter. We suggest some efficient estimation techniques which efficiently combine the information from both full model and submodel.

For estimation purposes, constraint on the parametric space can be presented in the form of the null hypothesis,

$$H_o : \gamma_1 = \gamma_2 = \cdots = \gamma_k = \gamma \text{ (unknown)}. \tag{3.3}$$

First, we propose a *candidate submodel estimator* (CSE) or so-called *pooled estimator* of the common parameter γ.

3.2.1 Candidate Submodel Estimation Strategy

A candidate submodel estimator or pooled estimator of γ is defined by

$$\hat{\boldsymbol{\gamma}}^{CS} = (\hat{\gamma}_n^{CS}, \ldots, \hat{\gamma}_n^{CS})' = \hat{\gamma}_n^{CS}\mathbf{1}_k, \quad \hat{\gamma}_n^{CS} = \sum_{i=1}^{k} n_i\hat{\gamma}_i/n, \quad n = n_1 + \cdots + n_k. \tag{3.4}$$

Similarly, the *shrinkage candidate submodel estimator* (SCSE) of γ may be defined as

$$\hat{\boldsymbol{\gamma}}^{SCS} = \hat{\boldsymbol{\gamma}} - \pi(\hat{\boldsymbol{\gamma}} - \hat{\boldsymbol{\gamma}}^{CS}), \quad \pi \in (0, 1), \tag{3.5}$$

where π is a constant and may be regarded as the degree of trust in the null hypothesis. If $\pi = 1$, then we obtain the CSE. Clearly, $\hat{\boldsymbol{\gamma}}^{SCS}$ is a convex combination of $\hat{\boldsymbol{\gamma}}$ and $\hat{\boldsymbol{\gamma}}^{CS}$ through a fixed value of $\pi \in (0, 1)$. As in the case of the one-sample problem (Chap. 2), we will show that both $\hat{\boldsymbol{\gamma}}^{SCS}$ and $\hat{\boldsymbol{\gamma}}^{CS}$ have a smaller MSE than $\hat{\boldsymbol{\gamma}}$ in an interval near the null hypothesis at the expense of its performance in the rest of the parameter space. Not only that, their MSEs become unbounded as the hypothesis error grows. If the prior information is *bad* in the sense that the hypothesis error is large, the pooled estimators are inferior to $\hat{\boldsymbol{\gamma}}$. Alternatively, if the information is *good*, i.e., the hypothesis error is small, $\hat{\boldsymbol{\gamma}}^{SCS}$ and $\hat{\boldsymbol{\gamma}}^{CS}$ offer a substantial MSE gain over $\hat{\boldsymbol{\gamma}}$.

The above insight leads to pretest and James–Stein type shrinkage estimation strategies when the hypothesis information is rather suspicious. A test statistic plays an integral role in construction of pretest and shrinkage estimators. For this reason we suggest following test statistic for the null hypothesis H_o in (3.3).

Test Statistic

A large-sample test statistic for the null hypothesis is obtained by defining the normalized distance of $\hat{\gamma}$ from $\hat{\gamma}^{CS}$:

$$\mathcal{D}_n = n(\hat{\boldsymbol{\gamma}} - \hat{\boldsymbol{\gamma}}^{CS})'\hat{\boldsymbol{\Gamma}}_n^{-1}(\hat{\boldsymbol{\gamma}} - \hat{\boldsymbol{\gamma}}^{CS}), \tag{3.6}$$

where

$$\hat{\boldsymbol{\Gamma}}_n^{-1} = \frac{\boldsymbol{\Omega}_n}{\hat{\tau}^2}, \quad \boldsymbol{\Omega}_n = \text{Diag}\left(\omega_{1,n}, \ldots, \omega_{k,n}\right), \quad \omega_{i,n} = \frac{n_i}{n}, \quad \hat{\tau}^2 = \frac{1}{2}(\hat{\gamma}^{CS})^2 + (\hat{\gamma}^{CS})^4.$$

Assuming $\lim(\omega_{i,n}) = \omega_i$ is fixed for $i = 1, \ldots, k$, and $\hat{\boldsymbol{\Gamma}}_n$ converges $\boldsymbol{\Gamma}$. Under the null hypothesis, the distribution of $\mathcal{D}_n$ converges in distribution to a χ^2 distribution with $(k-1)$ degrees of freedom. Hence, the upper α-level critical value of $\mathcal{D}_n$ defined by c_α may be approximated by this distribution. For more information, we refer to Ahmed (2002), and for $k = 1$ and 2, we refer to Ahmed (1994) and Ahmed (1995), respectively.

3.2.2 Pretest Estimation Strategy

The *pretest estimator* (PE) of $\boldsymbol{\gamma}$ is defined by

$$\hat{\boldsymbol{\gamma}}^{P} = \hat{\boldsymbol{\gamma}} - (\hat{\boldsymbol{\gamma}} - \hat{\boldsymbol{\gamma}}^{CS}) I(\mathscr{D}_n < c_\alpha), \tag{3.7}$$

where $I(A)$ is an indicator function of a set A. It is important to remark that $\hat{\boldsymbol{\gamma}}^{P}$ performs better than $\hat{\boldsymbol{\gamma}}$ in important parts of the parameter space.

Again, we remark here the use of $\hat{\boldsymbol{\gamma}}^{P}$ may, however, be limited due to the large size of the pretest. Further, we recall that $\hat{\boldsymbol{\gamma}}^{SCS}$ provides a wider range than $\hat{\boldsymbol{\gamma}}^{CS}$ in which it dominates $\hat{\boldsymbol{\gamma}}$. Thus, it is logical to replace $\hat{\boldsymbol{\gamma}}^{CS}$ by $\hat{\boldsymbol{\gamma}}^{SCS}$ in (3.7).

3.2.3 Shrinkage Pretest Estimation Strategy

The *shrinkage pretest estimator* (SPE) is defined by incorporating π in (3.7) or replacing $\hat{\boldsymbol{\gamma}}^{CS}$ by $\hat{\boldsymbol{\gamma}}^{SCS}$ in (3.7) as follows:

$$\hat{\boldsymbol{\gamma}}^{SP} = \hat{\boldsymbol{\gamma}} - \pi(\hat{\boldsymbol{\gamma}} - \hat{\boldsymbol{\gamma}}^{CS}) I(\mathscr{D}_n < c_\alpha). \tag{3.8}$$

In the two-sample problem ($k = 2$), Ahmed (1995) established that $\hat{\boldsymbol{\gamma}}^{SP}$ significantly improves upon $\hat{\boldsymbol{\gamma}}^{P}$ in the size of the test, and dominates $\hat{\boldsymbol{\gamma}}$ in a large portion of the parameter space.

It is well established in the reviewed literature that the estimators based on the pretest method are sensitive to departure from H_o and may not be efficient for all $\boldsymbol{\gamma}$. Thus, we propose Stein-type estimators which will combine the sample and nonsample information in a superior way to the preceding estimators.

3.2.4 Shrinkage Estimation Strategy

The *Stein-type shrinkage estimator* (SSE) is defined by

$$\hat{\boldsymbol{\gamma}}^{S} = \hat{\boldsymbol{\gamma}} - \{(k-3)\mathscr{D}_n^{-1}\}(\hat{\boldsymbol{\gamma}} - \hat{\boldsymbol{\gamma}}^{CS}), \quad k \geq 4. \tag{3.9}$$

By design and structure of this estimator, one can expect that this estimator will provide uniform improvement over $\hat{\boldsymbol{\gamma}}$, noting that suggested shrinkage estimator is not a convex combination of $\hat{\boldsymbol{\gamma}}^{CS}$ and $\hat{\boldsymbol{\gamma}}$. Therefore by design, the $\hat{\boldsymbol{\gamma}}^{S}$ may not remain nonnegative. In an effort to fix this problem with $\hat{\boldsymbol{\gamma}}^{S}$, we introduce a positive part shrinkage estimator.

3.2.5 Improved Shrinkage Estimation Strategy

We define the *positive part estimator (PSE)* as follows:

$$\begin{aligned}\hat{\gamma}^{PS} &= \hat{\gamma} - (k-3)\mathscr{D}_n^{-1}(\hat{\gamma} - \hat{\gamma}^{CS}) \\ &\quad - \{1-(k-3)\mathscr{D}_n^{-1}\}I(\mathscr{D}_n < k-3)(\hat{\gamma} - \hat{\gamma}^{CS}), \quad k \geq 4. \end{aligned} \tag{3.10}$$

Having defined all these estimators we need to assess the relative performance of these estimators. Accordingly, we establish an asymptotic criterion and establish some interesting and meaningful results in the section below to achieve our goal.

3.3 Asymptotic Theory and Methodology

Let us consider the following weighted quadratic loss function:

$$L(\gamma^*, \gamma) = n(\gamma^* - \gamma)' \boldsymbol{Q}(\gamma^* - \gamma), \tag{3.11}$$

where $\boldsymbol{\gamma}^*$ is an estimator of $\boldsymbol{\gamma}$ and $\boldsymbol{Q}$ is a known positive semi-definite matrix.

We plan to use the notion of asymptotic distributional risk to establish the asymptotic properties of all aforementioned estimators. To do so, let us assume that we have $G(\boldsymbol{y}) = \lim_{n\to\infty} P\{\sqrt{n}(\gamma^* - \gamma) \leq \boldsymbol{y}\}$. Now, we define the asymptotic distributional risk by

$$ADR(\gamma^*, \boldsymbol{Q}) \equiv R(\gamma^*, \boldsymbol{Q}) = \int\int\cdots\int \boldsymbol{y}'\boldsymbol{Q}\boldsymbol{y}dG(\boldsymbol{y}) = \text{trace}(\boldsymbol{Q}\boldsymbol{Q}^*), \tag{3.12}$$

where $\boldsymbol{Q}^* = \int\int\cdots\int \boldsymbol{y}\boldsymbol{y}'dG(\boldsymbol{y})$.

It is exceedingly important to calculate the ADR of the estimators when the null hypothesis may not hold. To be fair, we need to provide the general risk analysis of the suggested estimators relative to full model estimator when the null hypothesis or assumed parametric restriction may not hold. We achieve this objective by first defining a sequence of local alternatives as follows:

$$K_{(n)} : \boldsymbol{\gamma} = \boldsymbol{\gamma}_n, \quad \text{where} \quad \boldsymbol{\gamma}_n = \boldsymbol{\gamma} + \frac{\boldsymbol{\delta}}{\sqrt{n}}, \ \boldsymbol{\delta} \text{ is a real fixed vector.} \tag{3.13}$$

Note that that (3.3) is a special case of $\{K_{(n)}\}$.

Further, the test statistic in (3.6) is consistent against fixed $\boldsymbol{\gamma}$ such that $\boldsymbol{\gamma} \notin H_o$. Hence, both pretest and shrinkage estimators, involving the test statistic are asymptotically equivalent to the unrestricted estimator for the fixed alternatives. This further strengthened the use of contagious local alternative of our work. We refer to Ahmed (2001) for some insights on this matter.

Theorem 3.1 *If $\gamma \notin H_o$, then $\hat{\gamma}^{CS}$ and $\hat{\gamma}^{SCS}$ will have an unbounded asymptotic risk. However, $\hat{\gamma}^{P}$, $\hat{\gamma}^{SP}$, $\hat{\gamma}^{S}$, and $\hat{\gamma}^{PS}$ will have the same finite risk as of that of $\hat{\gamma}$.*

Proof A sketch of the proof is given below.

$$\begin{aligned}\sqrt{n}(\hat{\gamma}^{P}-\hat{\gamma})'\boldsymbol{Q}\sqrt{n}(\hat{\gamma}^{P}-\hat{\gamma}) &= \{I(\mathcal{D}_n < c_\alpha)\}\{\sqrt{n}(\hat{\gamma}-\hat{\gamma}^{CS})'\boldsymbol{Q}\sqrt{n}(\hat{\gamma}-\hat{\gamma}^{CS})\}\\ &\leq \{\mathcal{D}_n I(\mathcal{D}_n < c_\alpha)\}ch_{\max}(\boldsymbol{Q}\boldsymbol{\Omega}^{-1})\\ &\leq \{\mathcal{D}_n I(\mathcal{D}_n < c_\alpha)\}trace(\boldsymbol{Q}\boldsymbol{\Omega}^{-1}), \qquad (3.14)\end{aligned}$$

where $ch_{\max}(\boldsymbol{A})$ is the largest characteristic root of a matrix $\boldsymbol{A}$. For $\gamma \notin H_o$, $E\{\mathcal{D}_n I(\mathcal{D}_n < c_\alpha)\} \leq c_\alpha\{P(\mathcal{D}_n < c_\alpha)\}$. The test statistic $\mathcal{D}_n$ is consistent, hence $E\{\mathcal{D}_n I(\mathcal{D}_n < c_\alpha)\} \to 0$ as $n \to \infty$. Consequentially, for a fixed γ the estimators based on full model, $\hat{\gamma}^{P}$ and $\hat{\gamma}^{SP}$ will have the same ADR.

Investigating the asymptotic characteristic of $\hat{\gamma}^{S}$, we note that

$$\begin{aligned}\sqrt{n}(\hat{\gamma}^{S}-\hat{\gamma})'\boldsymbol{Q}\sqrt{n}(\hat{\gamma}^{S}-\hat{\gamma}) &= (k-3)^2\mathcal{D}_n^{-2}\{\sqrt{n}(\hat{\gamma}-\hat{\gamma}^{CS})'\boldsymbol{Q}\sqrt{n}(\hat{\gamma}-\hat{\gamma}^{CS})\}\\ &\leq (k-3)^2\{n(\hat{\gamma}-\hat{\gamma}^{CS})'\boldsymbol{Q}(\hat{\gamma}-\hat{\gamma}^{CS})\}^{-1}\\ &\leq \{ch_{\max}(\boldsymbol{Q}\boldsymbol{\Omega}^{-1})\}^2 \leq \{trace(\boldsymbol{Q}\boldsymbol{\Omega}^{-1})\}^2. \qquad (3.15)\end{aligned}$$

Noting that, if $\{\mathcal{D}_n = 0\}$, then we will have $\hat{\gamma}^{S} = \hat{\gamma} = \hat{\gamma}^{CS}$. For $\gamma \notin H_o$,

$$E\{\mathcal{D}_n^{-1} I(\mathcal{D}_n > 0)\} \to 0 \quad \text{as} \quad n \to \infty.$$

This clearly suggests that $\hat{\gamma}^{S}$ and $\hat{\gamma}$ will become asymptotically risk equivalent for every γ not in H_o. We can expect a similar analysis for $\hat{\gamma}^{PS}$.

On the other hand, $n \to \infty$ the estimator based on a submodel,

$$(\hat{\gamma}^{CS}-\gamma) \stackrel{a.s.}{\to} a(\neq 0),$$

for any $\gamma \notin H_o$,

$$n(\hat{\gamma}^{CS}-\gamma)'\boldsymbol{Q}(\hat{\gamma}^{CS}-\gamma) \stackrel{P}{\to} +\infty, \quad \text{as} \quad n \to \infty.$$

This clearly indicates that for large n the ADR of $\hat{\gamma}^{CS}$, approaches $+\infty$ when $\gamma \notin H_o$.

In the light of the results of the above theorem, we present the expression for ADR under local alternatives, and then compare the respective performances of the estimators. We begin the process, by establishing three lemmas. The results of these lemmas will facilitate the derivation of the ADR of the estimators.

Lemma 3.1 *Let us define*

$$X_n = \sqrt{n}(\hat{\gamma}-\gamma_o), \quad Y_n = \sqrt{n}(\hat{\gamma}-\hat{\gamma}^{CS}),$$

then under the local alternatives, we obtain the following joint asymptotic distribution

$$\begin{pmatrix} X_n \\ Y_n \end{pmatrix} \sim N_{2k} \left\{ \begin{pmatrix} \delta \\ \delta^* \end{pmatrix}, \begin{pmatrix} \Gamma & A \\ A' & A \end{pmatrix} \right\} \quad as \quad n \to \infty, \tag{3.16}$$

where

$$\delta^* = H\delta, \quad H = I_k - J\Omega, \quad J = \mathbf{1}_k \mathbf{1}_k^{'}, \quad \Gamma = \lim(\hat{\Gamma}_n), \quad A = \Gamma H'$$

Lemma 3.2 *Define* $Z_n = \sqrt{n}(\hat{\gamma}^{CS} - \gamma_o)$*, then under the local alternatives the joint distribution of*

$$\begin{pmatrix} Z_n \\ Y_n \end{pmatrix} \sim N_{2k} \left\{ \begin{pmatrix} \mathbf{0} \\ \delta^* \end{pmatrix}, \begin{pmatrix} \tau^2 J & 0 \\ 0 & A \end{pmatrix} \right\} \quad as \quad n \to \infty. \tag{3.17}$$

Here we assume that $\omega'\delta = 0$*, where* $\omega = (\omega_1, \ldots, \omega_k)$.

Lemma 3.3 *The test statistic* $\mathcal{D}_n$ *follows a noncentral chi-square distribution with* $(k-1)$ *degrees of freedom and a noncentrality parameter*

$$\Delta = \delta^{*\prime} \Gamma^{-1} \delta^*, \qquad as \ n \to \infty. \tag{3.18}$$

Hence, under the null hypothesis for large n, $\mathcal{D}_n$ *will closely follow the chi-square distribution with* $(k-1)$ *degrees of freedom. For given* α*, the critical value of* $\mathcal{D}_n$ *may be approximated by* $\chi^2_{k-1,\alpha}$*, the upper* $100\alpha\%$ *point of the chi-square distribution with* $(k-1)$ *degrees of freedom.*

3.4 Asymptotic Properties

In this section, we provide expressions for the *asymptotic distributional bias (ADB)* and ADR of the estimators.

To start the work, let $\Psi_k(x\ ;\ \Delta)$ to describe the noncentral chi-square distribution function with noncentrality parameter Δ and k degrees of freedom. Further, $E\left(\chi_k^{-2m}(\Delta)\right) = \int_0^\infty x^{-2m} d\psi_k(x\ ;\ \Delta)$. Now, we define the bias of an estimator γ^* as $B(\gamma^*) = E\{\lim_{n\to\infty} \sqrt{n}(\gamma^* - \gamma_{(n)})\}$.

The expressions for the bias of the suggested estimators are given in the theorem below:

Theorem 3.2

$$B(\hat{\gamma}) = \mathbf{0},$$

$$B(\hat{\gamma}^{CS}) = -\delta^*,$$

$$B(\hat{\gamma}^{SCS}) = -\pi\delta^*,$$

$$B(\hat{\boldsymbol{\gamma}}^{P}) = -\boldsymbol{\delta}^{*}\psi_{k+1}(\chi^2_{k-1,\alpha};\Delta),$$

$$B(\hat{\boldsymbol{\gamma}}^{SP}) = -\pi\boldsymbol{\delta}^{*}\psi_{k+1}(\chi^2_{k-1,\alpha};\Delta),$$

$$B(\hat{\boldsymbol{\gamma}}^{S}) = -(k-3)\boldsymbol{\delta}^{*}E(\chi^{-2}_{k+1}(\Delta)),$$

$$B(\hat{\boldsymbol{\gamma}}^{PS}) = -\boldsymbol{\delta}^{*}\left[\psi_{k+1}(k-3;\Delta) + E\{\chi^{-2}_{k+1}(\Delta)I(D_n > k-3)\}\right].$$

Proof

$$\begin{aligned} B(\hat{\boldsymbol{\gamma}}^{CS}) &= \lim_{n\to\infty}\sqrt{n}E(\hat{\boldsymbol{\gamma}^{CS}} - \boldsymbol{\gamma}_n) \\ &= E(-\boldsymbol{Y}_n) \\ &= -\boldsymbol{\delta}^{*}. \end{aligned}$$

$$\begin{aligned} B(\hat{\boldsymbol{\gamma}}^{SCS}) &= \lim_{n\to\infty}\sqrt{n}E(\hat{\boldsymbol{\gamma}}^{SCS} - \boldsymbol{\gamma}_n) \\ &= \lim_{n\to\infty}\sqrt{n}E(\hat{\boldsymbol{\gamma}}_n - \pi(\hat{\boldsymbol{\gamma}}_n - \hat{\boldsymbol{\gamma}}^{CS}) - \boldsymbol{\gamma}_n) \\ &= \lim_{n\to\infty}\sqrt{n}E(\hat{\boldsymbol{\gamma}}_n - \pi(\hat{\boldsymbol{\gamma}}_n - \hat{\boldsymbol{\gamma}}^{CS}) - \boldsymbol{\gamma}_0 - \frac{\boldsymbol{\delta}}{\sqrt{n}}) \\ &= E(\boldsymbol{X}_n) - \pi E(\boldsymbol{Y}_n) - \boldsymbol{\delta} \\ &= \boldsymbol{\delta} - \pi\boldsymbol{\delta}^{*} - \boldsymbol{\delta} \\ &= -\pi\boldsymbol{\delta}^{*}. \end{aligned}$$

$$\begin{aligned} B(\hat{\boldsymbol{\gamma}}^{P}) &= \lim_{n\to\infty}\sqrt{n}E(\hat{\boldsymbol{\gamma}}^{P} - \boldsymbol{\gamma}_n) \\ &= \lim_{n\to\infty}\sqrt{n}E(\hat{\boldsymbol{\gamma}}_n - (\hat{\boldsymbol{\gamma}}_n - \hat{\boldsymbol{\gamma}}^{CS})I(\mathscr{D}_n < c_\alpha) - \boldsymbol{\gamma}_n) \\ &= \lim_{n\to\infty}\sqrt{n}E(\hat{\boldsymbol{\gamma}}_n - (\hat{\boldsymbol{\gamma}}_n - \hat{\boldsymbol{\gamma}}^{CS})I(\mathscr{D}_n < c_\alpha) - \boldsymbol{\gamma}_0 - \frac{\boldsymbol{\delta}}{\sqrt{n}}) \\ &= E(\boldsymbol{X}_n) - E(\boldsymbol{Y}_n I(\mathscr{D}_n < c_\alpha)) - \boldsymbol{\delta} \\ &= \boldsymbol{\delta} - \boldsymbol{\delta}^{*}\psi_{k+1}(\chi^2_{k-1,\alpha};\Delta) - \boldsymbol{\delta} \\ &= -\boldsymbol{\delta}^{*}\psi_{k+1}(\chi^2_{k-1,\alpha};\Delta). \end{aligned}$$

$$\begin{aligned} B(\hat{\boldsymbol{\gamma}}^{SP}) &= \lim_{n\to\infty}\sqrt{n}E(\hat{\boldsymbol{\gamma}}^{SP} - \boldsymbol{\gamma}_n) \\ &= \lim_{n\to\infty}\sqrt{n}E(\hat{\boldsymbol{\gamma}}_n - \pi(\hat{\boldsymbol{\gamma}}_n - \hat{\boldsymbol{\gamma}}^{CS})I(\mathscr{D}_n < c_\alpha) - \boldsymbol{\gamma}_n) \end{aligned}$$

$$
\begin{aligned}
&= \lim_{n\to\infty} \sqrt{n}E(\hat{\gamma}_n - \pi(\hat{\gamma}_n - \hat{\gamma}^{CS})I(\mathscr{D}_n < c_\alpha) - \gamma_0 - \frac{\delta}{\sqrt{n}}) \\
&= E(\boldsymbol{X}_n) - \pi E(\boldsymbol{Y}_n I(\mathscr{D}_n < c_\alpha)) - \boldsymbol{\delta} \\
&= \boldsymbol{\delta} - \pi\boldsymbol{\delta}^* \psi_{k+1}(\chi^2_{k-1,\alpha}; \Delta) - \boldsymbol{\delta} \\
&= -\pi\boldsymbol{\delta}^* \psi_{k+1}(\chi^2_{k-1,\alpha}; \Delta).
\end{aligned}
$$

$$
\begin{aligned}
\boldsymbol{B}(\hat{\boldsymbol{\gamma}}^S) &= \lim_{n\to\infty} \sqrt{n}E(\hat{\boldsymbol{\gamma}}^S - \gamma_n) \\
&= \lim_{n\to\infty} \sqrt{n}E(\hat{\gamma}_n - \{(k-3)\mathscr{D}_n^{-1}\}(\hat{\gamma}_n - \hat{\boldsymbol{\gamma}}^{CS}) - \gamma_n) \\
&= \lim_{n\to\infty} \sqrt{n}E(\hat{\gamma}_n - \{(k-3)\mathscr{D}_n^{-1}\}(\hat{\gamma}_n - \hat{\boldsymbol{\gamma}}^{CS}) - \gamma_0 - \frac{\delta}{\sqrt{n}}) \\
&= E(\boldsymbol{X}_n) - (k-3)E(\boldsymbol{Y}_n\mathscr{D}_n^{-1}) - \boldsymbol{\delta} \\
&= \boldsymbol{\delta} - (k-3)\boldsymbol{\delta}^* E(\chi^{-2}_{k+1}(\Delta)) - \boldsymbol{\delta} \\
&= -(k-3)\boldsymbol{\delta}^* E(\chi^{-2}_{k+1}(\Delta)).
\end{aligned}
$$

$$
\begin{aligned}
\boldsymbol{B}(\hat{\boldsymbol{\gamma}}^{PS}) &= \lim_{n\to\infty} \sqrt{n}E(\hat{\boldsymbol{\gamma}}^{PS} - \gamma_n) \\
&= \lim_{n\to\infty} \sqrt{n}E(\hat{\boldsymbol{\gamma}}^S + (1-(k-3)\mathscr{D}_n^{-1})(\hat{\boldsymbol{\gamma}}^{CS} - \hat{\gamma}_n)I(\mathscr{D}_n < k-3) - \gamma_n) \\
&= \lim_{n\to\infty} \sqrt{n}E(\hat{\boldsymbol{\gamma}}^S - \gamma_n - (1-(k-3)E(\boldsymbol{Y}_n\mathscr{D}_n^{-1}I(\mathscr{D}_n < k-3),
\end{aligned}
$$

and after some algebraic manipulation we get the desired result.

Now, we use the following transformation to obtain a measurable analysis of the bias functions:

$$
B^*(.) = [\boldsymbol{B}(\boldsymbol{\gamma}^*)]'\boldsymbol{\Gamma}^{-1}[\boldsymbol{B}(\boldsymbol{\gamma}^*)],
$$

we term $B^*(.)$ as the asymptotic quadratic bias of an estimator of the parameter vector $\boldsymbol{\gamma}$. The corollary below showcases the expression for the asymptotic quadratic bias of the estimators.

Corollary 3.1

$$B^*(\hat{\boldsymbol{\gamma}}) = 0,$$

$$B^*(\hat{\boldsymbol{\gamma}}^{CS}) = \Delta,$$

$$B^*(\hat{\boldsymbol{\gamma}}^{SCS}) = \pi^2\Delta,$$

$$B^*(\hat{\boldsymbol{\gamma}}^{P}) = \Delta[\psi_{k+1}(\chi^2_{k-1,\alpha}; \Delta)]^2,$$

$$B^*(\hat{\boldsymbol{\gamma}}^{SP}) = \pi^2\Delta[\psi_{k+1}(\chi^2_{k-1,\alpha}; \Delta)]^2,$$

$$B^*(\hat{\boldsymbol{\gamma}}^S) = (k-3)^2 \Delta [E(\chi_{k+1}^{-2}(\Delta))]^2,$$

$$B^*(\hat{\boldsymbol{\gamma}}^{PS}) = \Delta \left[(k-3)E(\chi_{k+1}^{-2}(\Delta)) - E\{[(k-3)\chi_{k+1}^{-2}(\Delta) - 1] I(\mathcal{D}_n < k-3)\} \right]^2.$$

3.4.1 Bias Analysis

Clearly, the bias functions of all the estimators except $\hat{\boldsymbol{\gamma}}$ are a function of Δ. Since Δ is a common parameter for the bias functions, it makes sense that we examine the respective characteristics of the quadratic bias of the estimators for different values Δ. Observing that the magnitude of bias of $\hat{\boldsymbol{\gamma}}^{CS}$ and $\hat{\boldsymbol{\gamma}}^{SCS}$ increases without a bound and tends to ∞ as $\Delta \to \infty$; however, the bias of $\hat{\boldsymbol{\gamma}}^{CS}$ approaches infinity faster than that of $\hat{\boldsymbol{\gamma}}^{SCS}$.

The bias function of both pretest estimators starts from 0, increases to a certain point, then decreases gradually to 0. Further, $B^*(\hat{\boldsymbol{\gamma}}^{SP}) = \pi B^*(\hat{\boldsymbol{\gamma}}^P) < B^*(\hat{\boldsymbol{\gamma}}^P)$ for $\pi \in (0, 1)$. Thus, $\hat{\boldsymbol{\gamma}}^{SP}$ has asymptotically less bias than that of $\hat{\boldsymbol{\gamma}}^P$ depending upon the value of π. Hence, π may be considered as a bias reduction factor in the pretest estimation.

By nature, the shrinkage estimators are biased in general. The bias of $\hat{\boldsymbol{\gamma}}^S$ starts from 0 at $\Delta = 0$, then increases to a certain point, then decreases toward 0, since $E(\chi_\nu^{-2}(\Delta))$ is a decreasing log-convex function of Δ. The behavior of $\hat{\boldsymbol{\gamma}}^{PS}$ is similar to $\hat{\boldsymbol{\gamma}}^S$; however, the bias curve of $\hat{\boldsymbol{\gamma}}^{PS}$ remains below the curve of $\hat{\boldsymbol{\gamma}}^S$ for all values of Δ. Figure 3.1 displays these features of the estimators.

The expressions for ADR or simply risk are given in the following theorem.

Theorem 3.3 *Under local alternatives, the ADRs of the estimators are given by*

$$R(\hat{\boldsymbol{\gamma}}; \boldsymbol{Q}) = tr(\boldsymbol{Q\Gamma}), \tag{3.19}$$

$$R(\hat{\boldsymbol{\gamma}}^{CS}; \boldsymbol{Q}) = tr(\boldsymbol{Q\Gamma}) - tr(\boldsymbol{QB}) + \boldsymbol{\delta}^{*'} \boldsymbol{Q} \boldsymbol{\delta}^* \tag{3.20}$$

$$\textit{Where } \boldsymbol{B} = \boldsymbol{\Gamma} - \tau_o^2 \boldsymbol{J}$$

$$R(\hat{\boldsymbol{\gamma}}^{SCS}; \boldsymbol{Q}) = tr(\boldsymbol{Q\Gamma}) + \pi(\pi - 2) tr(\boldsymbol{QB}) + \pi^2 \boldsymbol{\delta}^{*'} \boldsymbol{Q} \boldsymbol{\delta}^* \tag{3.21}$$

$$\begin{aligned} R(\hat{\boldsymbol{\gamma}}^P; \boldsymbol{Q}) &= tr(\boldsymbol{Q\Gamma}) - tr(\boldsymbol{QB}) \psi_{k+1}(\chi_{k-1,\alpha}^2; \Delta) \\ &\quad - 2tr(\boldsymbol{Q\delta\delta}' \boldsymbol{\Gamma}^{-1} \boldsymbol{B}')[\psi_{k+3}(\chi_{k-1,\alpha}^2; \Delta) - \psi_{k+1}(\chi_{k-1,\alpha}^2; \Delta)] \\ &\quad + \boldsymbol{\delta}^{*'} \boldsymbol{Q} \boldsymbol{\delta}^* \psi_{k+3}(\chi_{k-1,\alpha}^2; \Delta) \end{aligned} \tag{3.22}$$

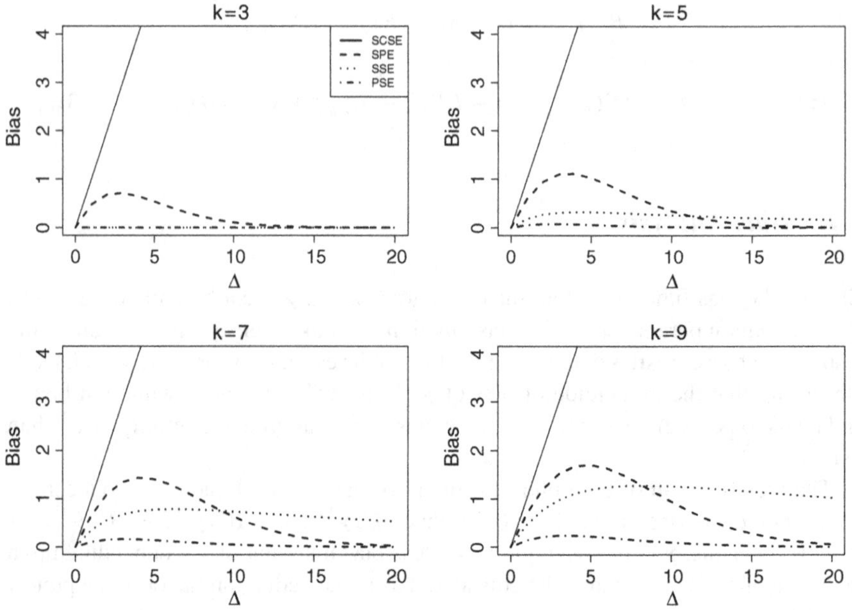

Fig. 3.1 Bias of the estimators for $k = 3, 5, 7, 9$, and $\alpha = 0.1, \pi = 1$

$$R(\hat{\boldsymbol{\gamma}}^{SP};\boldsymbol{Q}) = tr(\boldsymbol{Q\Gamma}) + \pi(\pi-2)tr(\boldsymbol{QB})\psi_{k+1}(\chi^2_{k-1,\alpha};\Delta) - 2\pi tr(\boldsymbol{Q\delta\delta'\Gamma}^{-1}\boldsymbol{B}')[\psi_{k+3}(\chi^2_{k-1,\alpha};\Delta) - \psi_{k+1}(\chi^2_{k-1,\alpha};\Delta)] + \pi^2\boldsymbol{\delta}^{*'}\boldsymbol{Q\delta}^{*}\psi_{k+3}(\chi^2_{k-1,\alpha};\Delta) \quad (3.23)$$

$$R(\hat{\boldsymbol{\gamma}}^{S};\boldsymbol{Q}) = tr(\boldsymbol{Q\Gamma}) - 2(k-3)tr(\boldsymbol{QB}')E(\chi^{-2}_{k+1}) - 2(k-3)tr(\boldsymbol{Q\delta\delta'\Gamma}^{-1}\boldsymbol{B}')[E(\chi^{-2}_{k+3}) - E(\chi^{-2}_{k+1})] + (k-3)^2 tr(\boldsymbol{QB})E(\chi^{-4}_{k+1}) + (k-3)^2\boldsymbol{\delta}^{*'}\boldsymbol{Q\delta}^{*}E(\chi^{-4}_{k+3}) \quad (3.24)$$

$$R(\hat{\boldsymbol{\gamma}}^{PS};\boldsymbol{Q}) = R(\hat{\boldsymbol{\gamma}}^{S};\boldsymbol{Q}) + 2tr(\boldsymbol{QB})E[(1-(k-3)\chi^{-2}_{k+1})I(\chi^2_{k+1} < (k-3))] - tr(\boldsymbol{QB})E[(1-(k-3)\chi^{-2}_{k+1})^2 I(\chi^2_{k+1} < (k-3))] + \boldsymbol{\delta}^{*'}\boldsymbol{Q\delta}^{*}E[(1-(k-3)\chi^{-2}_{k+3})^2 I(\chi^2_{k+1} < (k-3))]. \quad (3.25)$$

The proof of (3.19)–(3.25) is given below. First, we obtain the *mean squared error matrices* (MSEM) of the estimator, and then we obtain the risk expressions.

Proof

$$\begin{aligned}
MSEM(\hat{\boldsymbol{\gamma}}) &= E\left\{\lim_{n\to\infty} n(\hat{\boldsymbol{\gamma}} - \boldsymbol{\gamma}_n)(\hat{\boldsymbol{\gamma}} - \boldsymbol{\gamma}_n)'\right\}\\
&= E\left\{\lim_{n\to\infty} (\boldsymbol{X}_n - \boldsymbol{\delta})E(\boldsymbol{X}_n - \boldsymbol{\delta})'\right\}\\
&= \boldsymbol{\Gamma}.
\end{aligned}$$

Thus,

$$R(\hat{\boldsymbol{\gamma}}; \boldsymbol{Q}) = tr(\boldsymbol{Q\Gamma}).$$

$$\begin{aligned}
MSEM(\hat{\boldsymbol{\gamma}}^{CS}) &= E\left\{\lim_{n\to\infty} n(\hat{\boldsymbol{\gamma}}^{CS} - \boldsymbol{\gamma})(\hat{\boldsymbol{\gamma}}^{CS} - \boldsymbol{\gamma}_n)'\right\}\\
&= E\left\{\lim_{n\to\infty} n[\hat{\boldsymbol{\gamma}}^{CS} - \hat{\boldsymbol{\gamma}}_n + \hat{\boldsymbol{\gamma}} - \boldsymbol{\gamma}_n][\hat{\boldsymbol{\gamma}}^{CS} - \hat{\boldsymbol{\gamma}} + \hat{\boldsymbol{\gamma}}_n - \boldsymbol{\gamma}_n]'\right\}\\
&= E[-\boldsymbol{Y}_n + (\boldsymbol{X}_n - \boldsymbol{\delta})][-\boldsymbol{Y}_n + (\boldsymbol{X}_n - \boldsymbol{\delta})]'\\
&= E[\boldsymbol{Y}_n\boldsymbol{Y}'_n] - E[\boldsymbol{Y}_n(\boldsymbol{X}_n - \boldsymbol{\delta})']\\
&\quad - E[(\boldsymbol{X}_n - \boldsymbol{\delta})\boldsymbol{Y}'_n] + E[(\boldsymbol{X}_n - \boldsymbol{\delta})(\boldsymbol{X}_n - \boldsymbol{\delta})']\\
&= \boldsymbol{B} + \boldsymbol{\delta}^*\boldsymbol{\delta}^{*'} - 2E[\boldsymbol{X}_n\boldsymbol{Y}'_n] + 2\boldsymbol{\delta}E(\boldsymbol{Y}'_n) + \boldsymbol{\Gamma}\\
&= \boldsymbol{B} + \boldsymbol{\delta}^*\boldsymbol{\delta}^{*'} - 2\boldsymbol{B} + \boldsymbol{\Gamma}\\
&= \boldsymbol{\Gamma} - \boldsymbol{B} + \boldsymbol{\delta}^*\boldsymbol{\delta}^{*'}.
\end{aligned}$$

Hence,

$$R(\hat{\boldsymbol{\gamma}}^{CS}; \boldsymbol{Q}) = tr(\boldsymbol{Q\Gamma}) - tr(\boldsymbol{QB}) + \boldsymbol{\delta}^{*'}\boldsymbol{Q}\boldsymbol{\delta}^*.$$

$$\begin{aligned}
MSEM(\hat{\boldsymbol{\gamma}}^{SCS}) &= E\left\{\lim_{n\to\infty} n(\hat{\boldsymbol{\gamma}}^{SCS} - \boldsymbol{\gamma}_n)(\hat{\boldsymbol{\gamma}}^{SCS} - \boldsymbol{\gamma}_n)'\right\}\\
&= E\left\{\lim_{n\to\infty} n[\hat{\boldsymbol{\gamma}}_n - \boldsymbol{\gamma}_n - \pi(\hat{\boldsymbol{\gamma}}_n - \hat{\boldsymbol{\gamma}}^{CS})]\right.\\
&\qquad \left.[\hat{\boldsymbol{\gamma}}_n - \boldsymbol{\gamma}_n - \pi(\hat{\boldsymbol{\gamma}}_n - \hat{\boldsymbol{\gamma}}^{CS})]'\right\}\\
&= E[\hat{\boldsymbol{\gamma}}_n - \boldsymbol{\gamma}_n - \pi\boldsymbol{Y}_n][\hat{\boldsymbol{\gamma}}_n - \boldsymbol{\gamma}_n - \pi\boldsymbol{Y}_n]'\\
&= E[(\boldsymbol{X}_n - \boldsymbol{\delta}) - \pi\boldsymbol{Y}_n][(\boldsymbol{X}_n - \boldsymbol{\delta}) - \pi\boldsymbol{Y}_n]'\\
&= E[(\boldsymbol{X}_n - \boldsymbol{\delta})(\boldsymbol{X}_n - \boldsymbol{\delta})'] - \pi E[(\boldsymbol{X}_n - \boldsymbol{\delta})\boldsymbol{Y}'_n]\\
&\quad - \pi E[\boldsymbol{Y}_n(\boldsymbol{X}_n - \boldsymbol{\delta})'] + \pi^2 E[\boldsymbol{Y}_n\boldsymbol{Y}'_n]\\
&= \boldsymbol{\Gamma} - 2\pi E[\boldsymbol{X}_n\boldsymbol{Y}'_n] + 2\pi\boldsymbol{\delta}E(\boldsymbol{Y}'_n) + \pi^2 E[\boldsymbol{Y}_n\boldsymbol{Y}'_n]\\
&= \boldsymbol{\Gamma} - 2\pi\boldsymbol{B} + \pi^2[\boldsymbol{B} + \boldsymbol{\delta}^*\boldsymbol{\delta}^{*'}].
\end{aligned}$$

Therefore,

$$R(\hat{\boldsymbol{\gamma}}^{SCS}; \boldsymbol{Q}) = tr(\boldsymbol{Q}\boldsymbol{\Gamma}) + \pi(\pi - 2)tr(\boldsymbol{Q}\boldsymbol{B}) + \pi^2 \boldsymbol{\delta}^{*'} \boldsymbol{Q}\boldsymbol{\delta}^*.$$

$$\begin{aligned}
MSEM(\hat{\boldsymbol{\gamma}}^P) &= E\left\{\lim_{n\to\infty} n(\hat{\boldsymbol{\gamma}}^P - \boldsymbol{\gamma}_n)(\hat{\boldsymbol{\gamma}}^P - \boldsymbol{\gamma}_n)'\right\} \\
&= E\left\{\lim_{n\to\infty} n[\hat{\boldsymbol{\gamma}}_n - \boldsymbol{\gamma}_n - (\hat{\boldsymbol{\gamma}}_n - \hat{\boldsymbol{\gamma}}^{CS}) I(\mathscr{D}_n < k-3)]\right. \\
&\quad \left.[\hat{\boldsymbol{\gamma}}_n - \boldsymbol{\gamma}_n - (\hat{\boldsymbol{\gamma}}_n - \hat{\boldsymbol{\gamma}}^{CS}) I(\mathscr{D}_n < k-3)]'\right\} \\
&= E\left\{\lim_{n\to\infty} n[\hat{\boldsymbol{\gamma}}_n - \boldsymbol{\gamma}_n - \boldsymbol{Y}_n I(\mathscr{D}_n < k-3)]\right. \\
&\quad \left.[\hat{\boldsymbol{\gamma}}_n - \boldsymbol{\gamma}_n - \boldsymbol{Y}_n I(\mathscr{D}_n < k-3)]'\right\} \\
&= E[(\boldsymbol{X}_n - \boldsymbol{\delta}) - \boldsymbol{Y}_n I(\mathscr{D}_n < k-3)] \\
&\quad [(\boldsymbol{X}_n - \boldsymbol{\delta}) - \boldsymbol{Y}_n I(\mathscr{D}_n < k-3)]' \\
&= E[(\boldsymbol{X}_n - \boldsymbol{\delta})(\boldsymbol{X}_n - \boldsymbol{\delta})'] - E[(\boldsymbol{X}_n - \boldsymbol{\delta})\boldsymbol{Y}'_n I(\mathscr{D}_n < k-3)] \\
&\quad - E[\boldsymbol{Y}_n I(\mathscr{D}_n < k-3)(\boldsymbol{X}_n - \boldsymbol{\delta})'] + E[\boldsymbol{Y}_n \boldsymbol{Y}'_n I(\mathscr{D}_n < k-3)] \\
&= \boldsymbol{\Gamma} - 2E[\boldsymbol{X}_n \boldsymbol{Y}'_n I(\mathscr{D}_n < k-3)] + 2\boldsymbol{\delta} E(\boldsymbol{Y}'_n I(\mathscr{D}_n < k-3)) \\
&\quad + E[\boldsymbol{Y}_n \boldsymbol{Y}'_n I(\mathscr{D}_n < k-3)] \\
&= \boldsymbol{\Gamma} - 2\boldsymbol{B}' \psi_{k+1}(\chi^2_{k-1,\alpha}; \Delta) \\
&\quad - 2\boldsymbol{\delta}\boldsymbol{\delta}' \boldsymbol{\Gamma}^{-1} \boldsymbol{B}[\psi_{k+3}(\chi^2_{k-1,\alpha}; \Delta) - \psi_{k+1}(\chi^2_{k-1,\alpha}; \Delta)] \\
&\quad - 2\boldsymbol{\delta}\boldsymbol{\delta}^{*'} \psi_{k+1}(\chi^2_{k-1,\alpha}; \Delta) + 2\boldsymbol{\delta}\boldsymbol{\delta}^{*'} \psi_{k+1}(\chi^2_{k-1,\alpha}; \Delta) \\
&\quad + \boldsymbol{B}\psi_{k+1}(\chi^2_{k-1,\alpha}; \Delta) + \boldsymbol{\delta}^*\boldsymbol{\delta}^{*'} \psi_{k+3}(\chi^2_{k-1,\alpha}; \Delta) \\
&= \boldsymbol{\Gamma} - \boldsymbol{B}\psi_{k+1}(\chi^2_{k-1,\alpha}; \Delta) \\
&\quad - 2\boldsymbol{\delta}\boldsymbol{\delta}' \boldsymbol{\Gamma}^{-1} \boldsymbol{B}'[\psi_{k+3}(\chi^2_{k-1,\alpha}; \Delta) - \psi_{k+1}(\chi^2_{k-1,\alpha}; \Delta)] \\
&\quad + \boldsymbol{\delta}^*\boldsymbol{\delta}^{*'} \psi_{k+3}(\chi^2_{k-1,\alpha}; \Delta).
\end{aligned}$$

Thus,

$$\begin{aligned}
R(\hat{\boldsymbol{\gamma}}^P; \boldsymbol{Q}) &= tr(\boldsymbol{Q}\boldsymbol{\Gamma}) - tr(\boldsymbol{Q}\boldsymbol{B})\psi_{k+1}(\chi^2_{k-1,\alpha}; \Delta) \\
&\quad - 2tr(\boldsymbol{Q}\boldsymbol{\delta}\boldsymbol{\delta}' \boldsymbol{\Gamma}^{-1} \boldsymbol{B}')[\psi_{k+3}(\chi^2_{k-1,\alpha}; \Delta) - \psi_{k+1}(\chi^2_{k-1,\alpha}; \Delta)] \\
&\quad + \boldsymbol{\delta}^{*'} \boldsymbol{Q}\boldsymbol{\delta}^* \psi_{k+3}(\chi^2_{k-1,\alpha}; \Delta).
\end{aligned}$$

$$\begin{aligned}
MSEM(\hat{\boldsymbol{\gamma}}^{SP}) &= E\left\{\lim_{n\to\infty} n(\hat{\boldsymbol{\gamma}}^{SP} - \boldsymbol{\gamma}_n)(\hat{\boldsymbol{\gamma}}^{SP} - \boldsymbol{\gamma}_n)'\right\} \\
&= E\left\{\lim_{n\to\infty} n[\hat{\boldsymbol{\gamma}}_n - \boldsymbol{\gamma}_n - \pi(\hat{\boldsymbol{\gamma}}_n - \hat{\boldsymbol{\gamma}}^{CS}) I(\mathscr{D}_n < k-3)]\right. \\
&\quad \left.[\hat{\boldsymbol{\gamma}}_n - \boldsymbol{\gamma}_n - \pi(\hat{\boldsymbol{\gamma}}_n - \hat{\boldsymbol{\gamma}}^{CS}) I(\mathscr{D}_n < k-3)]'\right\}
\end{aligned}$$

$$
\begin{aligned}
&= E\left\{\lim_{n\to\infty} n[\hat{\boldsymbol{\gamma}}_n - \boldsymbol{\gamma}_n - \pi \boldsymbol{Y}_n I(\mathscr{D}_n < k-3)]\right.\\
&\quad \left.[\hat{\boldsymbol{\gamma}}_n - \boldsymbol{\gamma}_n - \pi \boldsymbol{Y}_n I(\mathscr{D}_n < k-3)]'\right\}\\
&= E\left\{\lim_{n\to\infty} [(\boldsymbol{X}_n - \boldsymbol{\delta}) - \pi \boldsymbol{Y}_n I(\mathscr{D}_n < k-3)]\right.\\
&\quad \left.[(\boldsymbol{X}_n - \boldsymbol{\delta}) - \pi \boldsymbol{Y}_n I(\mathscr{D}_n < k-3)]'\right\}\\
&= E[(\boldsymbol{X}_n - \boldsymbol{\delta})(\boldsymbol{X}_n - \boldsymbol{\delta})'] - \pi E[(\boldsymbol{X}_n - \boldsymbol{\delta})\boldsymbol{Y}'_n I(\mathscr{D}_n < k-3)]\\
&\quad - \pi E[\boldsymbol{Y}_n I(\mathscr{D}_n < k-3)(\boldsymbol{X}_n - \boldsymbol{\delta})'] + \pi^2 E[\boldsymbol{Y}_n \boldsymbol{Y}'_n I(\mathscr{D}_n < k-3)]\\
&= \boldsymbol{\Gamma} - 2\pi E[\boldsymbol{X}_n \boldsymbol{Y}'_n I(\mathscr{D}_n < k-3)] + 2\pi \boldsymbol{\delta} E(\boldsymbol{Y}'_n I(\mathscr{D}_n < k-3))\\
&\quad + \pi^2 E[\boldsymbol{Y}_n \boldsymbol{Y}'_n I(\mathscr{D}_n < k-3)]\\
&= \boldsymbol{\Gamma} - 2\pi \boldsymbol{B}' \psi_{k+1}(\chi^2_{k-1,\alpha}; \Delta)\\
&\quad - 2\pi \boldsymbol{\delta}\boldsymbol{\delta}' \boldsymbol{\Gamma}^{-1} \boldsymbol{B}[\psi_{k+3}(\chi^2_{k-1,\alpha}; \Delta) - \psi_{k+1}(\chi^2_{k-1,\alpha}; \Delta)]\\
&\quad - 2\pi \boldsymbol{\delta}\boldsymbol{\delta}^{*'} \psi_{k+1}(\chi^2_{k-1,\alpha}; \Delta) + 2\pi \boldsymbol{\delta}\boldsymbol{\delta}^{*'} \psi_{k+1}(\chi^2_{k-1,\alpha}; \Delta)\\
&\quad + \pi^2 \boldsymbol{B} \psi_{k+1}(\chi^2_{k-1,\alpha}; \Delta) + \pi^2 \boldsymbol{\delta}^* \boldsymbol{\delta}^{*'} \psi_{k+3}(\chi^2_{k-1,\alpha}; \Delta)\\
&= \boldsymbol{\Gamma} - 2\pi \boldsymbol{B} \psi_{k+1}(\chi^2_{k-1,\alpha}; \Delta)\\
&\quad - 2\pi \boldsymbol{\delta}\boldsymbol{\delta}' \boldsymbol{\Gamma}^{-1} \boldsymbol{B}'[\psi_{k+3}(\chi^2_{k-1,\alpha}; \Delta) - \psi_{k+1}(\chi^2_{k-1,\alpha}; \Delta)]\\
&\quad + \pi^2 \boldsymbol{B} \psi_{k+1}(\chi^2_{k-1,\alpha}; \Delta) + \pi^2 \boldsymbol{\delta}^* \boldsymbol{\delta}^{*'} \psi_{k+3}(\chi^2_{k-1,\alpha}; \Delta).
\end{aligned}
$$

Hence,

$$
\begin{aligned}
R(\hat{\boldsymbol{\gamma}}^{SP}; \boldsymbol{Q}) &= tr(\boldsymbol{Q}\boldsymbol{\Gamma}) + \pi(\pi - 2) tr(\boldsymbol{Q}\boldsymbol{B}) \psi_{k+1}(\chi^2_{k-1,\alpha}; \Delta)\\
&\quad - 2\pi tr(\boldsymbol{Q}\boldsymbol{\delta}\boldsymbol{\delta}' \boldsymbol{\Gamma}^{-1} \boldsymbol{B}')[\psi_{k+3}(\chi^2_{k-1,\alpha}; \Delta) - \psi_{k+1}(\chi^2_{k-1,\alpha}; \Delta)]\\
&\quad + \pi^2 \boldsymbol{\delta}^{*'} \boldsymbol{Q} \boldsymbol{\delta}^* \psi_{k+3}(\chi^2_{k-1,\alpha}; \Delta).
\end{aligned}
$$

$$
\begin{aligned}
MSEM(\hat{\boldsymbol{\gamma}}^S) &= E\left\{\lim_{n\to\infty} n(\hat{\boldsymbol{\gamma}}^S - \boldsymbol{\gamma}_n)(\hat{\boldsymbol{\gamma}}^S - \boldsymbol{\gamma}_n)'\right\}\\
&= E\left\{\lim_{n\to\infty} n[\hat{\boldsymbol{\gamma}}_n - \boldsymbol{\gamma}_n - (k-3)\mathscr{D}_n^{-1}(\hat{\boldsymbol{\gamma}}_n - \hat{\boldsymbol{\gamma}}^{CS})]\right.\\
&\quad \left.[\hat{\boldsymbol{\gamma}}_n - \boldsymbol{\gamma}_n - (k-3)\mathscr{D}_n^{-1}(\hat{\boldsymbol{\gamma}}_n - \hat{\boldsymbol{\gamma}}^{CS})]'\right\}\\
&= E\left\{\lim_{n\to\infty} [(\boldsymbol{X}_n - \boldsymbol{\delta}) - (k-3)\mathscr{D}_n^{-1}\boldsymbol{Y}_n][(\boldsymbol{X}_n - \boldsymbol{\delta}) - (k-3)\mathscr{D}_n^{-1}\boldsymbol{Y}_n]'\right\}\\
&= E[(\boldsymbol{X}_n - \boldsymbol{\delta})(\boldsymbol{X}_n - \boldsymbol{\delta})'] - (k-3)E[(\boldsymbol{X}_n - \boldsymbol{\delta})\boldsymbol{Y}'_n \mathscr{D}_n^{-1}]\\
&\quad - (k-3)E[\mathscr{D}_n^{-1}\boldsymbol{Y}_n(\boldsymbol{X}_n - \boldsymbol{\delta})']\\
&\quad + (k-3)^2 E[\boldsymbol{Y}_n \boldsymbol{Y}'_n \mathscr{D}_n^{-2}]\\
&= \boldsymbol{\Gamma} - 2(k-3)E[\boldsymbol{X}_n \boldsymbol{Y}'_n \mathscr{D}_n^{-1}] + 2(k-3)\boldsymbol{\delta} E(\boldsymbol{Y}'_n \mathscr{D}_n^{-1})\\
&\quad + (k-3)^2 E[\boldsymbol{Y}_n \boldsymbol{Y}'_n \mathscr{D}_n^{-2}]\\
&= \boldsymbol{\Gamma} - 2(k-3)\boldsymbol{B}' E(\chi^{-2}_{k+1}) - 2(k-3)\boldsymbol{\delta}\boldsymbol{\delta}' \boldsymbol{\Gamma}^{-1} \boldsymbol{B}'[E(\chi^{-2}_{k+3}) - E(\chi^{-2}_{k+1})]\\
&\quad - 2(k-3)\boldsymbol{\delta}\boldsymbol{\delta}^{*'} E(\chi^{-2}_{k+1}) + 2(k-3)\boldsymbol{\delta}\boldsymbol{\delta}^{*'} E(\chi^{-2}_{k+1})
\end{aligned}
$$

$$
\begin{aligned}
&+(k-3)^2\boldsymbol{B}E(\chi_{k+1}^{-4})+(k-3)^2\boldsymbol{\delta}^*\boldsymbol{\delta}^{*'}E(\chi_{k+3}^{-4})\\
&=\boldsymbol{\Gamma}-2(k-3)\boldsymbol{B}'E(\chi_{k+1}^{-2})-2(k-3)\boldsymbol{\delta}\boldsymbol{\delta}'\boldsymbol{\Gamma}^{-1}\boldsymbol{B}'[E(\chi_{k+3}^{-2})-E(\chi_{k+1}^{-2})]\\
&+(k-3)^2\boldsymbol{B}E(\chi_{k+1}^{-4})+(k-3)^2\boldsymbol{\delta}^*\boldsymbol{\delta}^{*'}E(\chi_{k+3}^{-4}),
\end{aligned}
$$

so the risk of the estimator is

$$
\begin{aligned}
R(\hat{\boldsymbol{\gamma}}^S;\boldsymbol{Q})&=tr(\boldsymbol{Q}\boldsymbol{\Gamma})\\
&-2(k-3)tr(\boldsymbol{Q}\boldsymbol{B}')E(\chi_{k+1}^{-2})\\
&-2(k-3)tr(\boldsymbol{Q}\boldsymbol{\delta}\boldsymbol{\delta}'\boldsymbol{\Gamma}^{-1}\boldsymbol{B}')[E(\chi_{k+3}^{-2})-E(\chi_{k+1}^{-2})]\\
&+(k-3)^2tr(\boldsymbol{Q}\boldsymbol{B})E(\chi_{k+1}^{-4})+(k-3)^2\boldsymbol{\delta}^{*'}\boldsymbol{Q}\boldsymbol{\delta}^*E(\chi_{k+3}^{-4}).
\end{aligned}
$$

$$
\begin{aligned}
MSEM(\hat{\boldsymbol{\gamma}}^{PS})&=E\left\{\lim_{n\to\infty}n(\hat{\boldsymbol{\gamma}}^{PS}-\boldsymbol{\gamma}_n)(\hat{\boldsymbol{\gamma}}^{PS}-\boldsymbol{\gamma}_n)'\right\}\\
&=E\left\{\lim_{n\to\infty}n[\hat{\boldsymbol{\gamma}}^S-\boldsymbol{\gamma}_n-(1-(k-3)E(\boldsymbol{Y}_n\mathscr{D}_n^{-1}I(\mathscr{D}_n<k-3))]\right.\\
&\quad\left.[\hat{\boldsymbol{\gamma}}^S-\boldsymbol{\gamma}_n-(1-(k-3)E(\boldsymbol{Y}_n\mathscr{D}_n^{-1}I(\mathscr{D}_n<k-3))]'\right\}\\
&=MSEM(\hat{\boldsymbol{\gamma}}^S)-2E\{(\hat{\boldsymbol{\gamma}}^S-\boldsymbol{\gamma}_n)\boldsymbol{Y}'_n(1-(k-3)\mathscr{D}_n^{-1})I(\mathscr{D}_n<k-3)\}\\
&\quad+E(\boldsymbol{Y}_n\boldsymbol{Y}'_n(1-(k-3)\mathscr{D}_n^{-1})^2I(\mathscr{D}_n<k-3))\}\\
&=MSEM(\hat{\boldsymbol{\gamma}}^S)+2E(\boldsymbol{Y}_n\boldsymbol{Y}'_n(1-(k-3)\mathscr{D}_n^{-1})I(\mathscr{D}_n<k-3))\\
&\quad-2E(\boldsymbol{Y}_n\boldsymbol{Y}'_n(1-(k-3)\mathscr{D}_n^{-1})^2I(\mathscr{D}_n<k-3))\\
&\quad+E(\boldsymbol{Y}_n\boldsymbol{Y}'_n(1-(k-3)\mathscr{D}_n^{-1})^2I(\mathscr{D}_n<k-3))\\
&=MSEM(\hat{\boldsymbol{\gamma}}^S)+2E(\boldsymbol{Y}_n\boldsymbol{Y}'_n(1-(k-3)\mathscr{D}_n^{-1})I(\mathscr{D}_n<k-3))\\
&\quad-E(\boldsymbol{Y}_n\boldsymbol{Y}'_n(1-(k-3)\mathscr{D}_n^{-1})^2I(\mathscr{D}_n<k-3))\\
&=MSEM(\hat{\boldsymbol{\gamma}}^S)+2\boldsymbol{B}E[(1-(k-3)\chi_{k+1}^{-2})I(\chi_{k+1}^2<(k-3))]\\
&\quad+2\boldsymbol{\delta}^*\boldsymbol{\delta}^{*'}E[(1-(k-3)\chi_{k+3}^{-2})I(\chi_{k+1}^2<(k-3))]\\
&\quad-\boldsymbol{B}E[(1-(k-3)\chi_{k+1}^{-2})^2I(\chi_{k+1}^2<(k-3))]\\
&\quad-\boldsymbol{\delta}^*\boldsymbol{\delta}^{*'}E[(1-(k-3)\chi_{k+3}^{-2})^2I(\chi_{k+1}^2<(k-3))],
\end{aligned}
$$

where $E\{(\hat{\boldsymbol{\gamma}}^S-\boldsymbol{\gamma}_n)\boldsymbol{Y}'_n(1-(k-3)\mathscr{D}_n^{-1})I(\mathscr{D}_n<k-3)\}$ equals to

$$
\begin{aligned}
&E\{[(\hat{\boldsymbol{\gamma}}^{CS}-\boldsymbol{\gamma}_n)+(1-(k-3)\mathscr{D}_n^{-1})\boldsymbol{Y}_n]\boldsymbol{Y}'_n(1-(k-3)\mathscr{D}_n^{-1})I(\mathscr{D}_n<k-3)\}\\
&=E\{[-\boldsymbol{Y}_n+(1-(k-3)\mathscr{D}_n^{-1})\boldsymbol{Y}_n]\boldsymbol{Y}'_n(1-(k-3)\mathscr{D}_n^{-1})I(\mathscr{D}_n<k-3)\}\\
&=-E(\boldsymbol{Y}_n\boldsymbol{Y}'_n(1-(k-3)\mathscr{D}_n^{-1})I(\mathscr{D}_n<k-3))\\
&\quad+E(\boldsymbol{Y}_n\boldsymbol{Y}'_n(1-(k-3)\mathscr{D}_n^{-1})^2I(\mathscr{D}_n<k-3)).
\end{aligned}
$$

Finally, we obtain the risk expression as follows:

$$R(\hat{\boldsymbol{\gamma}}^{PS};\boldsymbol{Q}) = R(\hat{\boldsymbol{\gamma}}^{S};\boldsymbol{Q}) + 2tr(\boldsymbol{QB})E[(1-(k-3)\chi_{k+1}^{-2})I(\chi_{k+1}^{2} < (k-3))]$$
$$- 2tr(\boldsymbol{QB})E[(1-(k-3)\chi_{k+1}^{-2})^{2}I(\chi_{k+1}^{2} < (k-3))]$$
$$+ \boldsymbol{\delta}^{*'}\boldsymbol{Q}\boldsymbol{\delta}^{*}E[(1-(k-3)\chi_{k+3}^{-2})^{2}I(\chi_{k+1}^{2} < (k-3))].$$

3.4.2 Risk Analysis

We critically examine and discuss the risk functions of the estimators and investigate the relative performance of all the estimators.

First of all, the ADR of $\hat{\boldsymbol{\gamma}}$ is constant (independent of Δ) with the value trace($\boldsymbol{Q\Gamma}$). The risk functions of the other estimators involved a common parameter, Δ, so we mainly study the properties of the estimators using risk functions in terms of Δ for the comparison purpose.

The risk of $\hat{\boldsymbol{\gamma}}^{CS}$ is a linear function of Δ and it becomes unbounded as Δ increases. Noting that

$$R(\hat{\boldsymbol{\gamma}}^{CS};\boldsymbol{Q}) \le R(\hat{\boldsymbol{\gamma}};\boldsymbol{Q}) \text{ if } \Delta \le \boldsymbol{QB}.$$

The ADR of $\hat{\boldsymbol{\gamma}}^{SCS}$ has similar characteristics.

The pretest estimators have some interesting properties. The ADR of $\hat{\boldsymbol{\gamma}}^{SP}$ is bounded in Δ and it begins with an initial value of [trace($\boldsymbol{Q\Gamma}$) − trace($\boldsymbol{QB}$)$\pi(2-\pi)\psi_{k+1}(\chi_{k-1,\alpha}^{2};0)$]. The risk of pretest estimators is lowest at $\Delta = 0$. However, as Δ deviates from the null hypothesis, the ADR function of $\hat{\boldsymbol{\gamma}}^{SP}$ monotonically approaches the ADR of $\hat{\boldsymbol{\gamma}}$ after crossing the risk function of $\hat{\boldsymbol{\gamma}}$ and achieving a maximum value. The risk function of $\hat{\boldsymbol{\gamma}}^{P}$ also follows a similar pattern. In fact,

$$\frac{R(\hat{\boldsymbol{\gamma}}^{SP};\boldsymbol{Q})}{R(\hat{\boldsymbol{\gamma}}_{n}^{CS},\boldsymbol{Q})} \le 1 \text{ if}$$

$$\Delta_{k} \le \frac{\text{trace}(\boldsymbol{QB})\{1-\pi(2-\pi)\psi_{k+1}(\chi_{k-1,\alpha}^{2};\Delta)\}}{1-2\pi\psi_{k+1}(\chi_{k-1,\alpha}^{2};\Delta)+\pi(2-\pi)\psi_{k+3}(\chi_{k-1,\alpha}^{2};\Delta)}.$$

In other words, $\hat{\boldsymbol{\gamma}}^{SP}$ is superior to $\hat{\boldsymbol{\gamma}}^{CS}$ if

$$\Delta_{k} \in \left[0, \frac{\text{trace}(\boldsymbol{QB})\{1-\pi(2-\pi)\psi_{k+1}(\chi_{k-1,\alpha}^{2};\Delta)\}}{1-2\pi\psi_{k+1}(\chi_{k-1,\alpha}^{2};\Delta)+\pi(2-\pi)\psi_{k+3}(\chi_{k-1,\alpha}^{2};\Delta)}\right).$$

We note that

$$R(\hat{\boldsymbol{\gamma}}^{SP};\boldsymbol{Q}) \le R(\hat{\boldsymbol{\gamma}};\boldsymbol{Q}) \text{ if}$$

$$\Delta_k \leq \frac{\text{trace}(\boldsymbol{QB})(2-\pi)\psi_{k+1}(\chi^2_{k-1,\alpha};\Delta)}{2\psi_{k+1}(\chi^2_{k-1,\alpha};\Delta)-(2-\pi)\psi_{k+3}(\chi^2_{k-1,\alpha};\Delta)},$$

and

$$R(\hat{\boldsymbol{\gamma}}^P;\boldsymbol{Q}) \leq R(\hat{\boldsymbol{\gamma}};\boldsymbol{Q}) \quad \text{if}$$

$$\Delta_k \leq \frac{\text{trace}(\boldsymbol{QB})\psi_{k+1}(\chi^2_{k-1,\alpha};\Delta)}{2\psi_{k+1}(\chi^2_{k-1,\alpha};\Delta)-\psi_{k+3}(\chi^2_{k-1,\alpha}.\Delta)}.$$

Hence, $\hat{\boldsymbol{\gamma}}^{SP}$ has better performance in a bigger part of parameter space than $\hat{\boldsymbol{\gamma}}^P$. As a special case, $\hat{\boldsymbol{\gamma}}^{SP}$ dominates $\hat{\boldsymbol{\gamma}}$ in the interval $\left[0, \text{trace}(\boldsymbol{QB})(2-\pi)\pi^{-1}\right)$, and $\hat{\boldsymbol{\gamma}}^P$ performs better than $\hat{\boldsymbol{\gamma}}$ if $\Delta \in [0, \text{trace}\,\boldsymbol{QB})$. This shows the superiority of $\hat{\boldsymbol{\gamma}}^{SP}$ over $\hat{\boldsymbol{\gamma}}^P$.

Now, let us determine the conditions under which $\hat{\boldsymbol{\gamma}}^{SP}$ dominates $\hat{\boldsymbol{\gamma}}^P$. First, consider the case when $\Delta = 0$. In this case,

$$R(\hat{\boldsymbol{\gamma}}^{SP};\boldsymbol{Q}) - R(\hat{\boldsymbol{\gamma}}^P;\boldsymbol{Q}) = \text{trace}(\boldsymbol{QB})(1-\pi)^2\psi_{k+1}(\chi^2_{k-1,\alpha};\Delta) > 0.$$

Thus, $\hat{\boldsymbol{\gamma}}^P$ has a smaller risk than that of $\hat{\boldsymbol{\gamma}}^{SP}$ when $\Delta = 0$. Alternatively,

$$\begin{aligned} R(\hat{\boldsymbol{\gamma}}^P;\boldsymbol{Q}) = R(\hat{\boldsymbol{\gamma}}^{SP};\,\boldsymbol{Q}) &- \text{trace}(\boldsymbol{QB})(1-\pi)^2\psi_{k+3}(\chi^2_{k-1,\alpha};\Delta) \\ &+ \Delta_k\left\{2(1-\pi)\psi_{k+1}(\chi^2_{k-1,\alpha};\Delta) - (1-\pi)^2\psi_{k+3}(\chi^2_{k-1,\alpha};\Delta)\right\}, \end{aligned} \tag{3.26}$$

indicating that the risk of $\hat{\boldsymbol{\gamma}}^P$ will be smaller than $\hat{\boldsymbol{\gamma}}^{SP}$ for rather small values of Δ, which may be negligible for larger values of π. As Δ increases, then the difference in (3.26) becomes positive and $\hat{\boldsymbol{\gamma}}^{SP}$ dominates $\hat{\boldsymbol{\gamma}}^P$ in the rest of the parameter space. For a given π, let Δ_{k_π} be a point in the parameter space at which the risk of $\hat{\boldsymbol{\gamma}}^{SP}$ and $\hat{\boldsymbol{\gamma}}^P$ intersect. Then, for $\Delta \in (0, \Delta_{k_\pi}]$, $\hat{\boldsymbol{\gamma}}^P$ performs better than $\hat{\boldsymbol{\gamma}}^{SP}$, while for $\Delta_k \in (\Delta_{k_\pi}, \infty)$, $\hat{\boldsymbol{\gamma}}^{SP}$ dominates $\hat{\boldsymbol{\gamma}}^P$.

We now turn to investigate the properties of the Stein-type shrinkage estimators. At $\Delta = 0$, the risk difference is

$$R(\hat{\boldsymbol{\gamma}};\boldsymbol{Q}) - R(\hat{\boldsymbol{\gamma}}^S;\boldsymbol{Q}) = \text{trace}(\boldsymbol{QB})(k-3)E\{2\chi^{-2}_{k+1}(0) - (k-3)\chi^{-4}_{k+1}(0)\} > 0.$$

Hence, we safely conclude that the Stein-type shrinkage estimators dominate the estimator based on full model when the nonsample information is correct. Also, the maximum risk gain of $\hat{\boldsymbol{\gamma}}^S$ over $\hat{\boldsymbol{\gamma}}$ is achieved at this value of Δ. However, it is important and fair to examine the relative performance of shrinkage estimator for all possible values of Δ, that is, for $\Delta \in (0, \infty)$. For comparison purpose, let us characterize a class of positive semi-definite matrices by

$$\boldsymbol{Q}^D = \left\{\frac{\text{trace}(\boldsymbol{Q\Gamma})}{ch_{max}(\boldsymbol{Q\Gamma})} \geq \frac{k+1}{2}\right\} \tag{3.27}$$

where $ch_{max}(\cdot)$ means the largest eigenvalue of $(\cdot)$. In order to provide a meaningful comparison of the various estimators, we state the following theorem:

Theorem 3.4 *(Courant Theorem) If* **C** *and* **D** *are two positive semi-definite* $q \times q$ *matrices with* **D** *nonsingular, then*

$$ch_{\min}(\mathbf{CD}^{-1}) \leq \frac{\mathbf{x}'\mathbf{Cx}}{\mathbf{x}'\mathbf{Dx}} \leq ch_{\max}(\mathbf{CD}^{-1}),$$

where $ch_{\min}(\cdot)$ *and* $ch_{\max}(\cdot)$ *mean the smallest and largest eigenvalues of* $(\cdot)$*, respectively, and* **x** *is a column vector.*

We note that the above lower and upper bounds are equal to the infimum and supremum, respectively, of the ratio $\frac{\mathbf{x}'\mathbf{Cx}}{\mathbf{x}'\mathbf{Dx}}$ for $\mathbf{x} \neq \mathbf{0}$. Also, for $\mathbf{D} = \mathbf{I}$, the ratio is known as the Rayleigh quotient for matrix **C**. As an application of the Courant theorem, we get

$$ch_{min}(\boldsymbol{Q\Gamma}) \leq \frac{\boldsymbol{\delta}^{*\prime}\boldsymbol{Q\delta}^{*}}{\boldsymbol{\delta}^{*\prime}\boldsymbol{\Gamma}^{-1}\boldsymbol{\delta}^{*}} \leq ch_{max}(\boldsymbol{Q\Gamma}), \quad \text{for } \boldsymbol{\delta}^{*} \neq \mathbf{0} \text{ and } \boldsymbol{Q} \in \boldsymbol{Q}^{D}.$$

Using the class of matrices defined in (3.27), we safely conclude that all $\boldsymbol{\delta}^{*}$, $R(\hat{\boldsymbol{\gamma}}^{S}; \boldsymbol{\gamma}) \leq R(\hat{\boldsymbol{\gamma}}; \boldsymbol{\gamma})$ where strict inequality holds for some $\boldsymbol{\delta}^{*}$.

More specifically, the value of risk of $\hat{\boldsymbol{\gamma}}^{S}$ is 3 at $\Delta = 0$. The value of risk increases monotonically toward trace$(\boldsymbol{QB})$ as Δ moves away from 0. The risk of $\hat{\boldsymbol{\gamma}}^{S}$ is uniformly smaller than $\hat{\boldsymbol{\gamma}}$ for many values of Δ. The upper limit is attained when $\Delta \to \infty$. In summary, the shrinkage estimator is far superior than the full model estimator when the submodel is nearly correctly specified. In the case where the submodel is grossly incorrectly specified, which is the worst possible scenario, then the performance of the shrinkage estimator will be as good as full model estimator, a very powerful and unparallel property of shrinkage estimator.

We now compare $\hat{\boldsymbol{\gamma}}^{S}$ and $\hat{\boldsymbol{\gamma}}^{CS}$ at $\Delta = 0$, we get

$$R(\hat{\boldsymbol{\gamma}}^{S}, \boldsymbol{Q}) - R(\hat{\boldsymbol{\gamma}}^{CS}, \boldsymbol{Q}) = \text{trace}(\boldsymbol{QB}) - \frac{k-3}{k-1}\text{trace}(\boldsymbol{QB}) > 0. \tag{3.28}$$

If the submodel is true then the estimator based on the submodel is far superior than any existing estimator in the reviewed literature, including shrinkage estimator. Having said that, as Δ departs from the origin $E(\chi_{k+1}^{-4}(\Delta))$ decreases, $\hat{\boldsymbol{\gamma}}^{S}$ has a smaller risk than $\hat{\boldsymbol{\gamma}}^{CS}$. Generally speaking, $\hat{\boldsymbol{\gamma}}^{S}$ does not perform better than $\hat{\boldsymbol{\gamma}}^{CS}$ in a small interval near the origin, and as a result $\hat{\boldsymbol{\gamma}}^{S}$ dominates $\hat{\boldsymbol{\gamma}}^{CS}$ in the rest of the parameter space. Hence, $\hat{\boldsymbol{\gamma}}^{S}$ is the relatively more effective estimator and provides an efficient alternative to $\hat{\boldsymbol{\gamma}}^{CS}$. This clearly demonstrates that the submodel estimator may be inconsistent when the model is not true and the estimation accuracy of this estimator cannot be trusted.

Next, we compare $\hat{\boldsymbol{\gamma}}^S$ and $\hat{\boldsymbol{\gamma}}^{SP}$ when submodel is true, that is, $\Delta = 0$.

$$R(\hat{\boldsymbol{\gamma}}^S, \boldsymbol{Q}) = R(\hat{\boldsymbol{\gamma}}^{SP}, \boldsymbol{Q}) + \text{trace}(\boldsymbol{QB}) \left\{ \pi(2-\pi)\psi_{k+1}(\chi^2_{k-1,\alpha}; 0) - \frac{(k-3)}{(k-1)} \right\}$$
$$> R(\hat{\boldsymbol{\gamma}}^{SP}, \boldsymbol{Q}), \quad k > 3, \tag{3.29}$$

satisfying

$$\pi < 1 - \sqrt{1 - \frac{(k-3)}{(k-1)\psi_{k+1}(\chi^2_{k-1,\alpha}; 0)}}, \quad \text{and} \quad \psi_{k+1}(\chi^2_{k-1,\alpha}; 0) < \frac{(k-1)}{(k-3)} \tag{3.30}$$

otherwise $\hat{\boldsymbol{\gamma}}^S$ has a smaller risk. A comparison for $\hat{\boldsymbol{\gamma}}^P$ versus $\hat{\boldsymbol{\gamma}}^S$ can be easily obtained by using $\pi = 1$ in the above expression. The risk of $\hat{\boldsymbol{\gamma}}^S$ and $\hat{\boldsymbol{\gamma}}^{SP}$ intersecting at $\Delta_k = \Delta_{k_{\pi,\alpha}}$ occurs if the condition (3.30) is satisfied; otherwise there is no intersecting point in the parameter space. If $\Delta_k \in [0, \Delta_{k_{\pi,\alpha}})$, then $R(\hat{\boldsymbol{\gamma}}^{SP}, \boldsymbol{Q}) \le R(\hat{\boldsymbol{\gamma}}^S, \boldsymbol{Q})$ while for $\Delta_k \in [\Delta_{k_{\pi,\alpha}}, \infty)$, $R(\hat{\boldsymbol{\gamma}}^S, \boldsymbol{Q}) \le R(\hat{\boldsymbol{\gamma}}^{SP}, \boldsymbol{Q})$. If (3.30) is not satisfied then $R(\hat{\boldsymbol{\gamma}}^S, \boldsymbol{Q}) \le R(\hat{\boldsymbol{\gamma}}^{SP}, \boldsymbol{Q})$.

Note that the application of $\hat{\boldsymbol{\gamma}}^S$ is constrained by the requirement that $k \ge 4$. If $k < 4$, then $\hat{\boldsymbol{\gamma}}^{SP}$ may be a good choice over estimators based on full model and submodel estimators, respectively.

Comparing the risk performance of $\hat{\boldsymbol{\gamma}}^{PS}$ and $\hat{\boldsymbol{\gamma}}^S$, we observed from relation (3.25) that

$$\frac{R(\hat{\boldsymbol{\gamma}}^{PS}, \boldsymbol{Q})}{R(\hat{\boldsymbol{\gamma}}^S, \boldsymbol{Q})} \le 1, \quad \text{for all } \boldsymbol{\delta}^*,$$

with strict inequality for some $\boldsymbol{\delta}^*$, establishing the fact that $\hat{\boldsymbol{\gamma}}^{PS}$ dominates $\hat{\boldsymbol{\gamma}}^S$. Thus, $\hat{\boldsymbol{\gamma}}^{PS}$ is superior to $\hat{\boldsymbol{\gamma}}$. Clearly, $\hat{\boldsymbol{\gamma}}^{PS}$ is preferable over $\hat{\boldsymbol{\gamma}}^S$. We suggest that $\hat{\boldsymbol{\gamma}}^S$ should not be used in its own right, however it can be used as a tool to construct $\hat{\boldsymbol{\gamma}}^{PS}$.

We observed that the ADR of all the estimators is a function of matrices $\boldsymbol{Q}$ and $\boldsymbol{\Gamma}$. The ADR expression can be drastically simplified by considering the case $\boldsymbol{Q} = \boldsymbol{\Gamma}^{-1}$. Now substitute this choice of $\boldsymbol{Q}$ in ADR expressions in (3.19)–(3.25). The risk expressions are given in corollary below.

Corollary 3.2

$$R(\hat{\boldsymbol{\gamma}}, \boldsymbol{\Gamma}^{-1}) = k \tag{3.31}$$

$$R(\hat{\boldsymbol{\gamma}}^{CS}, \boldsymbol{\Gamma}^{-1}) = 1 + \Delta \tag{3.32}$$

$$R(\hat{\boldsymbol{\gamma}}^{SCS}, \boldsymbol{\Gamma}^{-1}) = k - \pi(2-\pi)(k-1) + \pi^2\Delta \tag{3.33}$$

$$R(\hat{\boldsymbol{\gamma}}^P, \boldsymbol{\Gamma}^{-1}) = k - (k-1)\psi_{k+1}(\chi^2_{k-1,\alpha}; \Delta)$$
$$+ \Delta\{2\psi_{k+1}(\chi^2_{k-1,\alpha}; \Delta) - \psi_{k+3}(\chi^2_{k-1,\alpha}; \Delta)\}, \tag{3.34}$$

$$R(\hat{\gamma}^{SP}, \boldsymbol{\Gamma}^{-1}) = k - \pi(2-\pi)(k-1)\psi_{k+1}(\chi^2_{k-1,\alpha}; \Delta) + \Delta\{2\pi\psi_{k+1}(\chi^2_{k-1,\alpha}; \Delta) - \pi(2-\pi)\psi_{k+3}(\chi^2_{k-1,\alpha}; \Delta)\}, \quad (3.35)$$

$$\begin{aligned} R(\hat{\gamma}^{S}, \boldsymbol{\Gamma}^{-1}) = k &- 2(k-3)\Delta E(\chi^{-2}_{k+3} - 2(k-3)\Delta E(\chi^{-2}_{k+3}) \\ &+ 2(k-3)E(\chi^{-2}_{k+1})(1-k+\Delta) \\ &- (k-1)(k-3)^2 E(\chi^{-4}_{k+1}) + (k-3)^2\Delta E(\chi^{-4}_{k+3}) \quad (3.36) \end{aligned}$$

$$\begin{aligned} R(\hat{\gamma}^{PS}, \boldsymbol{\Gamma}^{-1}) = R(\hat{\gamma}^{S} &+ 2(k-1)E\left[(1-(k-3)\chi^2_{k+1})I(\chi^2_{k+1} < (k-3))\right] \\ &- (k-1)E\left[(1-(k-3)\chi^{-2}_{k+1})^2 I(\chi^2_{k+1} < (k-3))\right] \\ &+ \Delta E\left[(1-(k-3)\chi^{-2}_{k+3})^2 I(\chi^2_{k+1} < (k-3))\right] \quad (3.37) \end{aligned}$$

Further, these expressions are useful for numerical study. The risk functions can be plotted for a quick visual comparison. We refer to Ahmed (2002) for a numerical study of the estimators. We obtain the same risk analysis using above conical form of the risk function. The numerical results strongly corroborate with the theoretical findings.

3.5 Simulation Study

In this section, we investigate the relative performance of suggested estimators on simulated data. The main purpose of this simulation is to examine the performance of estimators based on a large-sample methodology in moderate sample situations.

We conduct a rather small simulation study to examine the properties of the suggested estimators for moderate samples. We calculate the risk of $\hat{\gamma}(R_1)$, $\hat{\gamma}^{CS}(R_2)$, $\hat{\gamma}^{P}(R_3)$, $\hat{\gamma}^{S}(R_4)$, and $\hat{\gamma}^{PS}(R_5)$ based on simulated data.

We define the *simulated relative efficiency* (RE) of an estimator $\gamma^\star$ to another estimator $\gamma^\diamond$ by

$$RE(\gamma^\star : \gamma^\diamond) = \frac{R(\gamma^\diamond)}{R(\gamma^\star)},$$

where $R(\gamma^\diamond)$ and $R(\gamma^\star)$ are the simulated risks of the estimators $\gamma^\star$ and $\gamma^\diamond$, respectively. Thus, a value of RE greater than 1 indicates the degree of superiority of $\gamma^\star$ over $\gamma^\diamond$. The simulated efficiency of various proposed estimators relative to $\hat{\gamma}$ is given by

$$RE_l = \frac{R_1}{R_q}, \quad q = 2, 3, 4, 5 \quad l < q,$$

where R_1, R_2, R_3, R_4, and R_5 are the simulated risks of $\hat{\gamma}$, $\hat{\gamma}^{CS}$, $\hat{\gamma}^{P}$, $\hat{\gamma}^{S}$, and $\hat{\gamma}^{PS}$, respectively. We assume that the populations have a normal distribution, and random numbers are generated for the given k and n_i. We consider an equal sample size case. The samples are drawn from normal populations with $\gamma_i = 0.2$. First, we calculate the test statistic $\mathscr{D}_n$ and then pretest and shrinkage estimators are calculated. The distribution of the test statistic $\mathscr{D}_n$ is computed under the null hypothesis, that is, $H_o : \gamma_1 = \gamma_2 = \cdots = \gamma_k$. The simulation for each estimator is repeated 1,000 times. The simulation results are showcased in the following tables.

We intend to examine the property of the estimators under realistic scenarios. Assuming all the parameters are equal is too optimistic (given that the submodel is true) and perhaps misleading. To this end, we define the parameter $\Delta^\star = \sum_{i=1}^{k}(\gamma_i - \gamma_{o_i})^2$. Thus, $\Delta^\star = 0$ implies that the selected submodel is correctly specified and no further action is required. However, in practice one seldom knows how to select a submodel. It is important to investigate the properties of estimators when such a submodel may not hold its ground. In other words, all the coefficients of variation may not be equal and $\Delta^\star > 0$. In an effort to examine the performance of the estimators for $\Delta^\star > 0$, further samples are simulated from normal populations which assumed a shift to the right by an amount $\Delta^\star$ when $\boldsymbol{\gamma} \neq \boldsymbol{\gamma}_o$.

The efficiency of the various estimators is computed based on 1,000 simulations for choices of k and α. Tables 3.1, 3.2, 3.3, 3.4, 3.5, 3.6, and 3.7 provide the estimated relative efficiency for the various estimates over $\hat{\gamma}$ for different sample sizes. Not surprisingly, tables reveal that maximum efficiency of all the estimators relative to $\hat{\gamma}$ achieved at $\Delta^\star = 0$.

Table 3.1 Relative efficiency for $k = 4$, $\alpha = 0.05$, and $n_i = 50$

Δ^*	RE_1	RE_2	RE_3	RE_4
0.00	3.82	2.65	1.35	1.53
0.15	0.33	0.72	1.00	1.01
0.20	0.11	0.99	1.00	1.00
0.30	0.05	1.00	1.00	1.00

Table 3.2 Relative efficiency for $k = 4$, $\alpha = 0.05$, and $n_i = 30$

Δ^*	RE_1	RE_2	RE_3	RE_4
0.00	4.06	2.76	1.34	1.51
0.15	0.52	0.72	1.04	1.05
0.20	0.16	0.87	0.99	1.00
0.30	0.11	0.93	1.00	1.01

Table 3.3 Relative efficiency for $k = 4$, $\alpha = 0.05$, and $n_i = 20$

Δ^*	RE_1	RE_2	RE_3	RE_4
0.00	3.90	2.63	1.39	1.51
0.15	0.75	0.79	1.09	1.11
0.20	0.26	0.73	0.99	1.01
0.30	0.20	0.81	1.01	1.00

Table 3.4 Relative efficiency for $k = 10$, $\alpha = 0.05$, and $n_i = 50$

Δ^*	RE_1	RE_2	RE_3	RE_4
0.00	10.22	5.17	3.33	4.41
0.15	0.61	0.85	1.25	1.27
0.20	0.18	0.97	1.03	1.04
0.30	0.11	0.98	1.01	1.05

Table 3.5 Relative efficiency for $k = 10$, $\alpha = 0.05$, and $n_i = 30$

Δ^*	RE_1	RE_2	RE_3	RE_4
0.00	10.14	4.63	3.39	4.17
0.15	0.96	0.98	1.44	1.45
0.20	0.30	0.99	1.09	1.10
0.30	0.21	0.98	1.05	1.06

Table 3.6 Relative efficiency for $k = 10$, $\alpha = 0.05$, and $n_i = 20$

Δ^*	RE_1	RE_2	RE_3	RE_4
0.00	9.39	4.89	3.38	3.95
0.15	1.33	1.13	1.63	1.69
0.20	0.47	0.88	1.17	1.33
0.30	0.35	0.91	1.13	1.21

Table 3.7 Relative efficiency for $k = 4$, $\alpha = 0.30$, and $n_i = 35$

Δ^*	RE_1	RE_2	RE_3	RE_4
0.00	4.08	1.38	1.39	1.47
0.15	0.55	0.97	1.05	1.07
0.20	0.19	0.99	0.99	1.01
0.30	0.11	1.00	1./00	1.01

Simulation study strongly corroborates the theoretical results and demonstrates that $\hat{\gamma}^{CS}$ performs exceptionally well with the other three estimators when $\Delta^\star$ is close to zero. Alternatively, as $\Delta^\star$ moves away from the origin, the simulation result shows that the performance of the submodel estimator deteriorates uniformly. Hence, it is not a desirable estimation strategy. On the other hand, the performance of $\hat{\gamma}^P$ is stable for such departures, that is, it achieves a maximum efficiency at $\Delta^\star = 0$ which drops and then tends to the risk of $\hat{\gamma}$ from below. Further, the relative efficiency of $\hat{\gamma}^P$ is higher than that of $\hat{\gamma}^S$ and $\hat{\gamma}^{PS}$ when $\Delta^\star$ is close to origin. However, for larger values of $\Delta^\star$, the opposite conclusion holds.

More importantly, $\hat{\gamma}^S$ and $\hat{\gamma}^{PS}$ are superior to $\hat{\gamma}$ for all the values of $\Delta^\star$. Further, $\hat{\gamma}^{PS}$ is relatively more efficient than $\hat{\gamma}^S$ in the entire parameter space induced by $\Delta^\star$. In a nutshell, Tables 3.1, 3.2, 3.3, 3.4, 3.5, 3.6, and 3.7 reveal that as $\Delta^\star$ is close to 0, all the proposed estimators are highly efficient relative to $\hat{\gamma}$. Further, for larger values of $\Delta^\star$, the performance of the estimators is similar to the analysis of asymptotic provided in Sect. 3.4. The estimators based on Stein-rule are the clear winners.

We have also assessed the performance of $\hat{\gamma}^S$ and $\hat{\gamma}^{PS}$ relative to $\hat{\gamma}^P$ for the larger sizes of the test, α. Both shrinkage estimators outperform the pretest estimator for larger values of α for all $\Delta^\star$. The results are documented in Table 3.7. It is seen that for $\alpha \geq 0.35$, the proposed shrinkage estimators dominate $\hat{\gamma}^P$ for all values of $\Delta^\star$. Interestingly, the positive part estimator dominates the pretest estimator for even $\alpha = 0.25$.

Simulation for other values of k were also performed, providing similar results and patterns. However, for large values of k the relative efficiency of shrinkage estimators over $\hat{\gamma}$ is substantial. For a detailed simulation study, we refer to Ahmed (2002). For $k = 2$, we refer to Ahmed et al. (1998).

3.6 Chapter Summary

Pooling several data from various sources to provide an estimation strategy is an interesting but a challenging problem. Researchers and professionals are equally interested in combining several data into a single data set for inference purposes. The pooled estimator based on a single data set is highly efficient under the stringent condition of the homogeneity of the parameters of several models. If this condition is violated, the resulting pooled estimator is highly biased and inconsistent. We suggested some alternative estimation strategies when we have multi-sample data of similar characteristics. We explored and discussed efficient estimation strategies that are a combination of pooled estimators and individual estimators.

Specifically, we suggested pretest and shrinkage estimators when multiple samples are available to increase the estimation by incorporating the test statistic in the estimation process. We demonstrate, both analytically and numerically, that our suggested shrinkage estimators based on Stein-rule are highly efficient. They perform better than the benchmark estimator $\hat{\gamma}$ in the entire parameter space. The pretest estimators perform better than the classical estimator in an important part of the parameter space. However, they fail to dominate the benchmark when the assumption of equality of the parameters may not hold. More importantly, the risk function of pretest estimators are bounded and do not explode.

In closing, we would like to emphasize that the estimator based on pooled data or based on a submodel alone may not be consistent and related inference based on this estimator may be misleading. Shrinkage and pretest estimators are considered to be best suited to the situation. Both methods combine the full model estimators and submodel estimators via a test statistic and produced satisfactory results. Having said that, the pooled or submodel estimator is the best when the assumed constraint on the parameter space is judiciously correct.

These strategies have been implemented in different contexts in the reviewed literature. We refer to Ahmed et al. (2001, 2006a,b, 2010a,b,c, 2011a, 2012), Muttlak et al. (2011), Ahmed and Liu (2009), Buhamra et al. (2007), Ghori et al. (2006), Ahmed and Khan (2002), Ahmed (2000a,b, 1992, 1999, 1988, 2003, 2005), and other.

References

Ahmed, S. E. (1988). Large-sample estimation strategies for eigenvalues of a wishart matrix. *Metrika*, *47*, 35–45.

Ahmed, S. E. (1992). Large sample pooling procedure for correlation. *The Statistician (Journal of Royal Statistical Society, Series D)*, *41*, 415–428.

Ahmed, S. E. (1994). Improved estimation of the coefficient of variation. *Journal of Applied Statistics*, *21*, 565–573.

Ahmed, S. E. (1995). A pooling methodology for coefficient of variation. *Sankhya*, *B*(57), 57–75.

Ahmed, S. E. (1999). Simultaneous estimation of survivor functions in exponential lifetime models. *Journal of Statistical Computation and Simulation*, *63*, 235–261.

Ahmed, S. E. (2000a). Construction of improved estimators of multinomial proportions. *Communications in Statistics: Theory and Methods*, *29*, 1273–1291.

Ahmed, S. E. (2000b). Stein-type shrinkage quantile estimation. *Stochastic Analysis and Applications*, *18*, 475–492.

Ahmed, S. E. (2001). Shrinkage estimation of regression coefficients from censored data with multiple observations. In S. Ahmed & N. Reid (Eds.), *Empirical Bayes and Likelihood Inference, Lecture Notes in Statistics* (Vol. 148, pp. 103–120). New York: Springer.

Ahmed, S. E. (2002). Simultaneous estimation of coefficients of variation. *Journal of Statistical Planning and Inference*, *104*, 31–51.

Ahmed, S. E. (2003). Pooling component means: Making sense or folly. In H. Yanai et al. (Eds.), *New Developments in Psychometrics*, (pp. 443–449). Tokyo: Springer.

Ahmed, S. E. (2005). Assessing process capability index for nonnormal processes. *Journal of Statistical Planning and Inference*, *129*, 195–206.

Ahmed, S. E., An, L., & Nkrunziza, S. (2006a). Improving the estimation of eigenvectors under quadratic loss. *Calcutta Statistical Association Bulletin*, 58.

Ahmed, S. E., Bhoj, D. S., & Ahsanullah, M. (1998). A monte carlo study of robustness of pretest and shrinkage estimators in pooling coefficients of variation. *Biometrical Journal*, *40*, 737–751.

Ahmed, S. E., Chitsaz, S., & Fallahpour, S. (2010a). *International Encyclopaedia of Statistical Science*, chapter Shrinkage Preliminary Test Estimation. Berlin: Springer.

Ahmed, S. E., Ghori, R., Goria, M. N., & Hussain, A. A. (2006b). Merging ginis indices under quadratic loss. *Statistica & Applicazioni, IV*, 47–55.
Ahmed, S. E., Gupta, A. K., Khan, S. M., & Nicol, C. J. (2001). Simultaneous estimation of several intraclass correlation coefficients. *Annals of the Institute of Statistical Mathematics, 53*, 354–369.
Ahmed, S. E., Hussein, A., & Nkurunziza, S. (2010b). Robust inference strategy in the presence of measurement error. *Statistics and Probability Letters, 80*, 726–732.
Ahmed, S. E., & Khan, S. M. (2002). Using several data to structure efficient estimation of intraclass correlation coefficients. In Nishisato et al. (Eds.), *Measurement and Multivariate Analysis*. Tokyo: Springer
Ahmed, S. E., & Liu, S. (2009). Asymptotic theory of simultaneous estimation of poisson means. *Linear Algebra and its Applications, 430*, 2734–2748.
Ahmed, S. E., Muttlak, H., Al-Mutawa, & Saheh, M. (2011a). Stein-type estimation using ranked set sampling. *Journal of Statistical Computation and Simulation*. doi:10.1080/00949655.2011.583248
Ahmed, S. E., Omar, M. H., & Joarder, A. H. (2012). Stabilizing the performance of kurtosis estimator of multivariate data. *Communications in Statistics - Simulation and Computation, 41*, 1860–1871.
Ahmed, S. E., Quadir, T., & Nkurunziza, S. (2010c). *International encyclopaedia of statistical science*, chapter Optimal Shrinkage Estimation. Berlin: Springer.
Buhamra, S., Al-Kandarri, N., & Ahmed, S. E. (2007). Nonparametric inference strategies for the quantile functions under left truncation and right censoring. *Journal of Nonparametric Statistics, 19*, 189–198.
Ghori, R., Ahmed, S. E., & Hussein, A. A. (2006). *Probability and statistics: Application and challenges*, chapter Shrinkage Estimation of Gini index (pp. 234–245). Singapore: World Scientific Publication.
Muttlak, H., Ahmed, S. E., & Al-Momani, M. (2011). Shrinkage estimation in replicated median ranked set sampling. *Journal of Statistical Computation and Simulation, 80*, 1185–1196.

Chapter 4
Estimation Strategies in Multiple Regression Models

Abstract In this chapter we present various large sample estimation strategies in a classical multiple regression model for estimation of the regression coefficients. These strategies are motivated by Stein-rule and pretest estimation procedures. In the context of two competing regression models (the full model and the candidate submodel), we suggest an adaptive shrinkage estimation technique that shrinks the full model estimate in the direction of the submodel estimate. The estimator based on pretest principle is also considered. Further, we apply the penalty estimation strategy for both variable selection and parameters estimation. We investigate the properties of the suggested estimators analytically and numerically. We provide the relative performance of all the listed estimators with the estimators based on the full model, respectively. Our analytical and simulation studies reveal that the shrinkage estimation strategy outperforms the estimation based on the full model procedure. Further, based on our limited simulation study, shrinkage and pretest estimators outperform penalty estimators when there are many inactive covariates in the model. Finally, the suggested methodology is evaluated through application to a real prostate data.

Keywords Regression models · Pretest and shrinkage estimation · Penalty estimation · Asymptotic bias and risk · Simulation · Prostrate data

4.1 Introduction

In the past decades, it has been known that pretest and shrinkage estimation strategies yield estimators which are superior in terms of risk to the maximum likelihood estimator (MLE). Particularly, the shrinkage estimator outshines the MLE in the entire parameter space. Until recently, shrinkage estimators have only been used to a limited extent in empirical applications, perhaps this is due to the computational burden and competing alternative estimation strategies. Interestingly, with

S. E. Ahmed, *Penalty, Shrinkage and Pretest Strategies*, SpringerBriefs in Statistics,
DOI: 10.1007/978-3-319-03149-1_4, © The Author(s) 2014

improvements in computing capability, and clear advantages to the use of prior information in many applications, this scenario is rapidly changing. For example, in many cases the shrinkage estimation strategy has been applied in the real estate market, where appraisers' expert knowledge can be fruitful. A real estate expert's opinion, experience, and knowledge often give precise information regarding certain parameter values in a housing pricing model. On the other hand, the recent literature on variable selection places an emphasis on using the data from an experiment to find a candidate subspace that represents a sparsity pattern in the predictor space. In the next stage, researchers may consider this auxiliary information and choose either the full model or the candidate submodel for subsequent work. The approach in this chapter is inspired by the Stein-rule and pretest estimation procedures. The Stein-rule strategy suggests that efficient estimates can be obtained by shrinking the full model estimates in the direction of submodel estimates. For this reason, we implement adaptive estimates where the amount of shrinkage depends on an estimate of the importance of the information left out of the subspace. Efron and Morris (1972) and many others considered extensions of the Stein-rule method. We consider pretest and Stein's approach to multiple regression models for the regression parameters estimation. Further, LASSO, adaptive LASSO and other penalty methods have become popular procedures of variable selection and estimation in a host of statistical models. Tibshirani (1996) introduced the least absolute shrinkage and selection operator (LASSO). This strategy forces some estimates to be shrunk toward exactly zero, thus resulting in simultaneous parameter selection and then estimation. LASSO is a member of a wide class of the *absolute penalty estimation* (APE) family. The penalty estimation strategy performs well if the model at hand is sparse. However, sparsity is a strong assumption and may not be judiciously justified in many situations.

In the arena of statistical inference ascertaining, the appropriate statistical model-estimator for use in representing the sample data is an interesting and challenging problem. Our plan here is to consider estimation problems under potential linear restrictions on the regression parameters in a multiple regression model. In numerous studies, many predicting variables are collected and included in the initial model building stage. However, having too many predictors in the model may result in increasing the uncertainty of the prediction results. For this and other reasons, variable selection is an exceedingly important measure of statistical analysis. Clearly, parsimony and reliability of predictors are desirable notions of statistical models. To this end, available prior information can be effectively used in the variables selection stage. Generally speaking, one possible source of prior information consists of knowing which of the predictor variables are of main interest and which variables are nuisance variables (candidate confounders) that may not effect the analysis of the association between the response and the main predictors. In some settings, examples of candidate confounders are age or origin of subjects or materials, and stratification, such as laboratory effect. As stated earlier, another source of prior information is the knowledge of results from previous experiments that search for sparsity patterns. This information at hand can be used to suggest a candidate subspace. A classical approach for dealing with such prior information would be to test whether the regression coefficients of the candidate confounders are zero, or, more generally, whether the full vector of parameters is in a given subspace. By design, shrinking the full model estimator in the direction of the subspace leads to more efficient estimators when the shrinkage is adaptive and based on the estimated distance between the subspace and the full space. Alternative procedures that have good performance are based on the penalty estimation technique. The penalty estimation strategy provides automatic parameter selection and estimation. However, this strategy completely relies on the selected submodel and ignores the information from the full model, if any.

Since its inception, shrinkage estimation has received considerable attention from researchers. Since 1987, Ahmed, his co-researchers, and others have analytically demonstrated that shrinkage estimators outshine the classical estimators based on the full model. They showed asymptotic properties of shrinkage and pretest estimators using a quadratic loss function, and demonstrated their dominance over the full model estimators. For example, Ahmed (1997) gave a detailed description of shrinkage estimation, and discussed large sample estimation techniques in a regression model with non-normal errors. Khan and Ahmed (2003) considered the problem of estimating the coefficient vector of a classical regression model, and demonstrated analytically and numerically that the positive-part of Stein-rule estimator, and the improved preliminary test estimator dominate the usual Stein-type, and pretest estimators, respectively. Ahmed and Nicol (2012) considered various large sample estimation techniques in a nonlinear regression model. Non-parametric estimation of the location parameter vector when uncertain prior information about the regression parameters is available was considered by Ahmed and Saleh (1990). Ahmed and Basu (2000) investigated the properties of shrinkage and pretest estimators in the general vector autoregressive process.

In this chapter, we consider the application of shrinkage and pretest estimation to the multiple regression model when it is *a priori* suspected that the coefficients may be restricted to a subspace or some auxiliary information is available via the variables selection procedure. In any event, we derive the expressions for asymptotic distributional bias and asymptotic distribution risk for the listed estimators. We show that shrinkage estimators have superior performance in terms of bias and risk over other estimators considered. More importantly, the shrinkage estimator is uniformly more efficient than the full model estimator under some general conditions. In contrast, the performance of the submodel and pretest estimators lacks this property. The relative performance of shrinkage estimators with penalty estimators such as LASSO, adaptive LASSO, and Smoothly Clipped Absolute Deviation (SCAD) estimators is evaluated through a simulation study. Further, the estimation strategy is evaluated through application to a real prostate data set.

The rest of the chapter is organized as follows. The model and suggested estimators are introduced in Sect. 4.2. The estimators are defined in Sect. 4.3. The asymptotic properties of the listed estimators are presented in Sect. 6.3. The results of a simulation study that includes a comparison of penalty procedures are provided in Sect. 6.4, along with the application to real data. Finally, Sect. 6.6 contains the concluding remarks and sheds some light for future research.

4.2 The Model and Statement of the Problem

In the context of linear regression, consider a problem of predicting the mean response using a set of regressors. If it is *a priori* known or suspected that a subset of the regressors does not significantly contribute to predicting the mean response, a submodel or restricted model excluding these covariates may be sufficient for the purpose. While it is often true that submodel estimators can offer a substantial improvement, in terms of mean squared error (MSE) over the full model estimator, there is still a concern that estimators are less desirable to use when the *uncertain prior information* (UPI) or the *auxiliary information* (AI) is incorrect. The advantage of shrinkage strategy, is therefore, that UPI/AI is incorporated into estimation to the extent that it appears to be true, given sample information. For this obvious reason, we therefore view the use of shrinkage estimators as an attractive and optimal trade-off in the context of a host of applications. Also, any UPI/AE may be validated through a pretest (or pretesting), and, depending on the outcome of the pretest, may be incorporated into the model as a parametric restriction. Thus, a pretest estimator chooses between the full model and the submodel estimators depending on the outcome of the preliminary test. However, the properties of pretest estimators are different from the full model and submodel estimation, respectively.

Model and Classical Estimation

Consider the following regression model:

$$\boldsymbol{Y} = \boldsymbol{X\beta} + \boldsymbol{\varepsilon}, \tag{4.1}$$

where $\boldsymbol{Y} = (y_1, y_2, \ldots, y_n)'$ is a vector of responses, $\boldsymbol{X}$ is an $n \times p$ fixed design matrix, $\boldsymbol{\beta} = (\beta_1, \ldots, \beta_p)'$ is an unknown vector of parameters, $\boldsymbol{\varepsilon} = (\varepsilon_1, \varepsilon_2, \ldots, \varepsilon_n)'$ is the vector of unobservable random errors, and the superscript ($'$) denotes the transpose of a vector or matrix.

We do not make any distributional assumption about the errors except that $\boldsymbol{\varepsilon}$ has a cumulative distribution function $F(\boldsymbol{\varepsilon})$ with $E(\boldsymbol{\varepsilon}) = \boldsymbol{0}$, and $E(\boldsymbol{\varepsilon\varepsilon}') = \sigma^2 \boldsymbol{I}$, where σ^2 is finite. Now, we list two assumptions commonly known as the regularity conditions:

(i) $\max_{1 \le i \le n} \boldsymbol{x}_i'(\boldsymbol{X}'\boldsymbol{X})^{-1}\boldsymbol{x}_i \to 0$ as $n \to \infty$, where $\boldsymbol{x}_i'$ is the ith row of $\boldsymbol{X}$, and

(ii) $\lim_{n \to \infty} \left(\frac{\boldsymbol{X}'\boldsymbol{X}}{n} \right) = \boldsymbol{C}$, where $\boldsymbol{C}$ is a finite positive-definite matrix.

The full model or unrestricted estimator (UE) of $\boldsymbol{\beta}$ is given by

$$\hat{\boldsymbol{\beta}}^{\text{UE}} = (\boldsymbol{X}'\boldsymbol{X})^{-1}\boldsymbol{X}'\boldsymbol{Y}.$$

Suppose the regression coefficients are restricted to a subspace as follows:

$$\boldsymbol{H\beta} = \boldsymbol{h},$$

where, $\boldsymbol{H}$ is a known matrix and $\boldsymbol{h}$ is a vector of known constants.

Thus, under the subspace restriction the submodel estimator or restricted estimator (RE) is given by

$$\hat{\boldsymbol{\beta}}^{\mathrm{RE}} = \hat{\boldsymbol{\beta}}^{\mathrm{UE}} - (\boldsymbol{X}'\boldsymbol{X})^{-1}\boldsymbol{H}'(\boldsymbol{H}(\boldsymbol{X}'\boldsymbol{X})^{-1}\boldsymbol{H}')^{-1}(\boldsymbol{H}\hat{\boldsymbol{\beta}}^{\mathrm{UE}} - \boldsymbol{h}).$$

A more interesting application of the above restriction is that $\boldsymbol{\beta}$ can be partitioned as $\boldsymbol{\beta} = (\boldsymbol{\beta}_1', \boldsymbol{\beta}_2')'$. The sub-vectors $\boldsymbol{\beta}_1$ and $\boldsymbol{\beta}_2$ are assumed to have dimensions p_1 and p_2 respectively, with $p_1 + p_2 = p$. In many applications we are interested in the estimation of $\boldsymbol{\beta}_1$ when it is plausible that $\boldsymbol{\beta}_2$ is a set of nuisance covariates. This situation may arise when there is over-modeling, and one wishes to cut down the irrelevant part from the model (4.1). In high-dimensional data analysis it is assumed that the model is sparse. In other words, it is plausible that $\boldsymbol{\beta}_2$ is near some specified $\boldsymbol{\beta}_2^o$, which, without loss of generality, may be set to a null vector.

In pretest estimation framework, we consider testing the restriction in the form of the following null hypothesis:

$$H_0 : \boldsymbol{H}\boldsymbol{\beta} = \boldsymbol{h}.$$

A suitable test statistic to test the above hypothesis is given by

$$\phi_n = \frac{(\boldsymbol{H}\hat{\boldsymbol{\beta}}^{\mathrm{UE}} - \boldsymbol{h})'(\boldsymbol{H}\boldsymbol{C}^{-1}\boldsymbol{H}')^{-1}(\boldsymbol{H}\hat{\boldsymbol{\beta}}^{\mathrm{UE}} - \boldsymbol{h})}{s_e^2}, \tag{4.2}$$

where

$$s_e^2 = \frac{(\boldsymbol{Y} - \boldsymbol{X}\hat{\boldsymbol{\beta}}^{\mathrm{UE}})'(\boldsymbol{Y} - \boldsymbol{X}\hat{\boldsymbol{\beta}}^{\mathrm{UE}})}{n-p}$$

is an estimator of σ^2. Under the null hypothesis the test statistic will follow a central chi-square distribution.

4.3 Estimation Strategies

We present shrinkage, pretest, and penalty estimation strategies in this section.

4.3.1 Shrinkage Estimation

We define a Stein-type shrinkage estimator (SE) $\hat{\boldsymbol{\beta}}_1^{\mathrm{S}}$ of $\boldsymbol{\beta}_1$ by

$$\hat{\boldsymbol{\beta}}_1^{\mathrm{S}} = \hat{\boldsymbol{\beta}}_1^{\mathrm{RE}} + (\hat{\boldsymbol{\beta}}_1^{\mathrm{UE}} - \hat{\boldsymbol{\beta}}_1^{\mathrm{RE}})\left\{1 - k\phi_n^{-1}\right\}, \text{ where } k = p_2 - 2, \quad p_2 \geq 3,$$

where ϕ_n is defined in (4.2).

A well-known issue with the SE is its tendency to "over-shrink" the resulting estimator beyond the full model estimator resulting in reversing the sign of the full model estimator. This could happen if $(p_2-1)\phi_n^{-1}$ is larger than one in absolute value. From the practical point of view, the change of sign would effect its interpretability. However, this behavior does not adversely effect the risk performance of the SE. To overcome the sign problem and moderate this effect, the *positive-rule Stein-type estimator* (PSE) has been suggested in the literature. A PSE has the form

$$\hat{\boldsymbol{\beta}}_1^{\mathrm{S+}} = \hat{\boldsymbol{\beta}}_1^{\mathrm{RE}} + (\hat{\boldsymbol{\beta}}_1^{\mathrm{UE}} - \hat{\boldsymbol{\beta}}_1^{\mathrm{RE}})\left\{1 - k\phi_n^{-1}\right\}^+, \text{ where } k = p_2 - 2, \quad p_2 \geq 3,$$

where we define the notation $z^+ = \max(0, z)$. This adjustment controls the over-shrinking problem in SE. Alternatively, PSE can be written in the following conical form:

$$\hat{\boldsymbol{\beta}}_1^{\mathrm{S+}} = \hat{\boldsymbol{\beta}}_1^{\mathrm{RE}} + (\hat{\boldsymbol{\beta}}_1^{\mathrm{UE}} - \hat{\boldsymbol{\beta}}_1^{\mathrm{RE}})\left\{1 - k\phi_n^{-1}\right\} I(\phi_n < k), \quad p_2 \geq 3.$$

Ahmed (2001), Ahmed and Chitsaz (2011), and Ahmed et al. (2012) have studied the asymptotic properties of Stein-type estimators in various contexts.

4.3.2 Pretest Estimation

A natural estimator is to define the *pretest estimator* (PE) or *preliminary test estimator* for the regression parameters $\boldsymbol{\beta}_1$ as

$$\hat{\boldsymbol{\beta}}_1^{\mathrm{PE}} = \hat{\boldsymbol{\beta}}_1^{\mathrm{UE}} - (\hat{\boldsymbol{\beta}}_1^{\mathrm{UE}} - \hat{\boldsymbol{\beta}}_1^{\mathrm{RE}}) I(\phi_n < c_{n,\alpha}), \tag{4.3}$$

where $I(\cdot)$ is an indicator function, and $c_{n,\alpha}$ is the $100(1-\alpha)$ percentage point of the test statistic ϕ_n.

Clearly, the pretest estimator is the full model estimator when the test statistic lies in a rejection region, and takes the value of the submodel estimator otherwise. Keep in mind that in a pretest estimation problem the UPI is tested before choosing the estimator for practical purposes, while a shrinkage and positive shrinkage estimator incorporates whatever UPI is available in the estimation process. A shrinkage estimator may be a smoothed version of the pretest estimator, since it is smooth function of the test statistics, ϕ_n.

4.3.3 Penalty Estimation

Penalty estimators are a class of estimators in the penalized least squares family of estimators, see Ahmed et al. (2010). This method involves penalizing the regression coefficients, and shrinking a subset of them to zero. In other words, the penalized procedure produces a submodel and subsequently estimates the submodel parameters. Several penalty estimators have been proposed in the literature for linear and generalized linear models. In this section, we consider the least absolute shrinkage and selection operation (LASSO) (Tibshirani 1996), the smoothly clipped absolute deviation method (SCAD) (Fan and Li 2001), the adaptive LASSO (Zou 2006), and minimax concave penalty (MCP) (Zhang 2010). By shrinking some regression coefficients to zero, these methods select parameters and estimation simultaneously.

Frank and Friedman (1993) introduced bridge regression, a generalized version of penalty (or absolute penalty type) estimators. For a given penalty function $\pi(\cdot)$ and regularization parameter λ, the general form can be written as

$$S(\boldsymbol{\beta}) = (\boldsymbol{y} - \boldsymbol{X}\boldsymbol{\beta})'(\boldsymbol{y} - \boldsymbol{X}\boldsymbol{\beta}) + \lambda\pi(\boldsymbol{\beta}),$$

where the penalty function is of the form

$$\pi(\beta) = \sum_{j=1}^{m} |\beta_j|^{\gamma}, \ \gamma > 0. \tag{4.4}$$

The penalty function in (4.4) bounds the L_{γ} norm of the parameters in the given model as $\sum_{j=1}^{m} |\beta_j|^{\gamma} \leq t$, where t is the tuning parameter that controls the amount of shrinkage. We see that for $\gamma = 2$, we obtain ridge estimates which are obtained by minimizing the penalized residual sum of squares

$$\hat{\boldsymbol{\beta}}^{\text{ridge}} = \underset{\boldsymbol{\beta}}{\operatorname{argmin}} \left\{ \sum_{i=1}^{n} (y_i - \beta_0 - \sum_{j=1}^{p} x_{ij}\beta_j)^2 + \lambda \sum_{j=1}^{p} \beta_j^2 \right\}, \tag{4.5}$$

where λ is the tuning parameter which controls the amount of shrinkage.

Frank and Friedman (1993) did not solve for the bridge regression estimators for any $\gamma > 0$. Interestingly, for $\gamma < 2$, it shrinks the coefficient toward zero, and depending on the value of λ, it sets some of them to be exactly zero. Thus, the procedure combines variable selection and shrinking of the coefficients of penalized regression.

An important member of the penalized least squares family is the L_1 penalized least squares estimator, which is obtained when $\gamma = 1$. This is known as the LASSO.

LASSO

LASSO was proposed by Tibshirani (1996), which performs variable selection and parameter estimation simultaneously. LASSO is closely related to ridge regression. LASSO solutions are similarly defined by replacing the squared penalty $\sum_{j=1}^{p} \beta_j^2$ in the ridge solution (4.5) with the absolute penalty $\sum_{j=1}^{p} |\beta_j|$ in the LASSO,

$$\hat{\boldsymbol{\beta}}^{\text{lasso}} = \underset{\boldsymbol{\beta}}{\operatorname{argmin}} \left\{ \sum_{i=1}^{n} (y_i - \beta_0 - \sum_{j=1}^{p} x_{ij}\beta_j)^2 + \lambda \sum_{j=1}^{p} |\beta_j| \right\}. \tag{4.6}$$

Although the change apparently looks subtle, the absolute penalty term made it impossible to have an analytic solution for the LASSO. Originally, LASSO solutions were obtained via quadratic programming. Later, Efron et al. (2004) proposed Least Angle Regression (LAR), a type of stepwise regression, with which the LASSO estimates can be obtained at the same computational cost as that of an ordinary least squares estimation. Further, the LASSO estimator remains numerically feasible for dimensions of p that are much higher than the sample size n. Friedman et al. (2010) developed an efficient algorithm for the estimation of a generalized linear model with a convex penalty, which efficiently computes the solution at a given regularization parameter. Thus, the whole process is repeated for typically 100 different regularization parameters to construct a piecewise linear approximation of the true nonlinear solution path.

Ahmed et al. (2007) proposed a penalty estimator for partially linear models. Further, they reappraised the properties of shrinkage estimators based on the Stein-rule estimation for the same model.

SCAD

Although the LASSO method does both shrinkage and variable selection due to the nature of the constraint region which often results in several coefficients becoming identically zero, it does not possess oracle properties (Fan and Li 2001). To overcome the inefficiency of traditional variable selection procedures, Fan and Li (2001) proposed SCAD to select variables and estimate the coefficients of variables automatically and simultaneously. This method not only retains the good features of both subset selection and ridge regression, but also produces sparse solutions, ensures continuity of the selected models (for the stability of model selection), and has unbiased estimates for large coefficients. The estimates are obtained as

$$\hat{\boldsymbol{\beta}}^{\text{SCAD}} = \underset{\boldsymbol{\beta}}{\operatorname{argmin}} \left\{ \sum_{i=1}^{n} (y_i - \beta_0 - \sum_{j=1}^{p} x_{ij}\beta_j)^2 + \lambda \sum_{j=1}^{p} p_{\alpha,\lambda} |\beta_j| \right\}.$$

Here $p_{\alpha,\lambda}(\cdot)$ is the smoothly clipped absolute deviation penalty. The solution of SCAD penalty is originally due to Fan (1997). SCAD penalty is a symmetric and a quadratic spline on $[0, \infty)$ with knots at λ and $\alpha\lambda$, whose first order derivative is given by

$$p_{\alpha,\lambda}(x) = \lambda \left\{ I(|x| \leq \lambda) + \frac{(\alpha\lambda - |x|)_+}{(\alpha - 1)\lambda} I(|x| > \lambda) \right\}, \quad x \geq 0. \tag{4.7}$$

Here $\lambda > 0$ and $\alpha > 2$ are the tuning parameters. For $\alpha = \infty$, the expression (4.7) is equivalent to the L_1 penalty.

Adaptive LASSO

Zou (2006) modified the LASSO penalty by using adaptive weights on L_1 penalties on the regression coefficients. Such a modified method was referred to as adaptive LASSO. It has been shown theoretically that the adaptive LASSO estimator is able to identify the true model consistently, and the resulting estimator is as efficient as the oracle.

The adaptive LASSO estimators (aLASSO) $\hat{\beta}^{\text{aLASSO}}$ are obtained by

$$\hat{\boldsymbol{\beta}}^{\text{aLASSO}} = \underset{\boldsymbol{\beta}}{\operatorname{argmin}} \left\{ \sum_{i=1}^{n} (y_i - \beta_0 - \sum_{j=1}^{p} x_{ij}\beta_j)^2 + \lambda \sum_{j=1}^{p} \hat{w}_j |\beta_j| \right\}, \tag{4.8}$$

where the weight function is

$$\hat{w}_j = \frac{1}{|\hat{\beta}_j^*|^\gamma}; \quad \gamma > 0,$$

and $\hat{\beta}_j^*$ is a root-n consistent estimator of β. Equation (4.8) is a "convex optimization problem and its global minimizer can be efficiently solved" (Zou 2006).

For example, $\hat{\beta}_j^*$ can be the ordinary least squares (OLS) estimator. Once we have the OLS estimator, we need to select $\gamma > 0$ and calculate the weights. Finally, the aLASSO estimates are obtained from (4.8). The LARS algorithm (Efron et al. 2004) can be used to obtain adaptive LASSO estimates. The steps are given below.

Step 1. Reweight the data by defining $\boldsymbol{x}_j^{\text{new}} = \boldsymbol{x}_j^{\text{old}}/\hat{w}_j, \quad j = 1, 2, \ldots, p$
Step 2. Solve the LASSO problem as

$$\hat{\boldsymbol{\beta}}^{**} = \underset{\boldsymbol{\beta}}{\operatorname{argmin}} \left\| \boldsymbol{y} - \textstyle\sum_{j=1}^{p} \boldsymbol{x}_j^{\text{new}} \beta_j \right\|^2 + \lambda \textstyle\sum_{j=1}^{p} |\beta_j|$$

Step 3. Return $\hat{\beta}_j^{\text{alasso}} = \hat{\beta}_j^{**}/\hat{w}_j$

For a detailed discussion on the computation of adaptive LASSO, we refer to Zou (2006).

MCP

Zhang (2010) suggested a *minimax concave penalty* (MCP) estimator which is given by

$$\hat{\boldsymbol{\beta}}_n^{\text{MCP}} = \operatorname{argmin}\left\{\sum_{i=1}^{n}\left(y_i - \sum_{j=1}^{p_n} x_{ij}\beta_j\right)^2 + \sum_{j=1}^{p_n} \rho_\lambda(|\beta_j|, \gamma)\right\},$$

where $\rho_\lambda(\gamma)$ is the MCP penalty given by

$$\rho_\lambda(\gamma) = \int_0^t (\lambda - x/\gamma)_+ \, dx,$$

where $\gamma > 0$ is a regularization parameter.

The above methods have been extensively studied in the literature. A bulk of research is ongoing, and it is hard to keep track of all the interesting work in this area. It has been pointed out in the reviewing literature that penalty estimators are not efficient when the dimension p becomes extremely large compared with sample size n. There are still challenging problems when p grows at a non-polynomial rate with n. Furthermore, non-polynomial dimensionality poses substantial computational challenges.

Raheem and Ahmed (2012) have studied some penalty estimators in linear regression models for fixed dimension, and have compared their predictive performance with shrinkage estimators. They observed that even in the case when p is fixed the performance of theses estimators is not superior when compared with the shrinkage estimators in some cases. The developments in the arena of penalty estimation are still in their infancy.

4.4 Submodel Selection

In the following, we explain how to obtain auxiliary information when dealing with a real data set to construct shrinkage and pretest estimators. For the data set at hand, we fit linear regression models to predict the variable of interest from the available regressors.

In the shrinkage and pretest estimation, we utilize the full model and submodel estimates and combine them in a way that shrinks the least squares estimates toward the sub-model estimates. If available in this framework, we utilize the information contained in the restricted subspace if it contributes significantly in predicting the response. However, in the absence of UPI about the nuisance subset, one might do a usual variable selection to filter the nuisance subset out of the covariates in the full model. To do this, one initiates the process with the model having all the covariates. Then, the best subset may be selected based on AIC, BIC, or some other model selection criteria. Separate estimates from the full model and a given submodel are then combined to obtain shrinkage or pretest estimates, respectively. Finally, a model with shrunken coefficients is obtained, which reduces the overall prediction error. The performance of each pair of full model and submodel is evaluated by estimating the prediction errors based on K-fold cross validation. In a cross validation, the data set is randomly divided into K subsets of roughly equal size. One subset is left aside and termed as test data, while the remaining $K-1$ subsets, called training sets, are used to fit the model. The fitted model is then used to predict the responses of the test data set. Finally, prediction errors are obtained by taking the squared deviation of the observed and predicted values in the test set.

We consider $K = 5, 10$. Both a raw cross-validation estimate (CVE), and a bias corrected cross-validation estimate of prediction errors are obtained for each configuration. The bias corrected cross-validation estimate is the adjusted cross-validation estimate designed to compensate for the bias introduced by not using the leave-one-out cross validation (Tibshirani and Tibshirani 2009).

Since cross-validation is a random process, the estimated prediction error varies across runs and for different values of K. To account for the random variation, we repeat the cross-validation process 5,000 times, and estimate the average prediction errors along with their standard errors. The number of repetitions was initially varied, and we settled with this as no noticeable variations in the standard errors were observed for higher values.

4.4.1 Prostate Data

Hastie et al. (2009) demonstrated various model selection techniques by fitting a linear regression model to the prostate data. Specifically, the log of prostate-specific antigen (`lpsa`) was modeled by the log cancer volume (`lcavol`), log prostate weight (`lweight`), age (`age`), log benign prostatic hyperplasia amount (`lbph`), seminal vesicle invasion (`svi`), log capsular penetration (`lcp`), Gleason score (`gleason`), and percentage Gleason scores 4 or 5 (`pgg45`). The idea is to predict `lpsa` from the measured variables.

The predictors were first standardized to have zero mean and unit standard deviation before fitting the model. Several model selection criteria were tried—details of which may be found in Hastie et al. (2009, Table 3.3, p. 63).

We consider the models obtained by AIC, BIC, and best subset selection (BSS) criteria, and consider them as our submodels. The three selection criteria do not give us the same submodels; however, most importantly, we obtain a submodel when given a full model, which allows us to construct shrinkage and pretest estimators. The submodels are listed in Table 4.1.

Table 4.1 Full and candidate submodels for prostate data

Selection criterion	Model: Response ~ Covariates
Full model	`lpsa ~ lcavol + lweight + svi + lbph + age + lcp + gleason + pgg45`
AIC	`lpsa ~ lcavol + lweight + svi + lbph + age`
BIC	`lpsa ~ lcavol + lweight + svi`
BSS	`lpsa ~ lcavol + lweight`

Now we turn our attention to shrinkage and pretest estimators, and we establish the asymptotic properties of these estimators in the following section.

4.5 Asymptotic Analysis

We examine the asymptotic properties of the full model, submodel, pretest, and shrinkage estimators when n is large and p is fixed. The goal is to derive the *asymptotic distributional bias* (ADB) and *asymptotic distributional risk* (ADR) of the estimator of $\boldsymbol{\beta}$. To achieve this goal, we consider a class of local alternatives $\{K_n\}$, which is given by

$$K_n : \boldsymbol{H\beta} = \boldsymbol{h} + \frac{\boldsymbol{\omega}}{\sqrt{n}}, \tag{4.9}$$

where $\boldsymbol{\omega} = (\omega_1, \omega_2, \ldots, \omega_{p_2})' \in R^{p_2}$ is a fixed vector. We notice that $\boldsymbol{\omega} = \boldsymbol{0}$ implies $\boldsymbol{H\beta} = \boldsymbol{h}$, i.e., the fixed alternative is a particular case of (4.9).

In the following, we evaluate the performance of each estimator under a local alternative. First, for an estimator $\hat{\boldsymbol{\beta}}^*$ and a positive-definite matrix $\boldsymbol{W}$, we define the weighted quadratic loss function

$$L(\hat{\boldsymbol{\beta}}^*; \boldsymbol{\beta}) = n(\hat{\boldsymbol{\beta}}^* - \boldsymbol{\beta})'\boldsymbol{W}(\hat{\boldsymbol{\beta}}^* - \boldsymbol{\beta}),$$

where $\boldsymbol{W}$ is the weighting matrix. For $\boldsymbol{W} = \boldsymbol{I}$, it is the simple squared error loss function.

The the risk function is defined as

$$E[L(\hat{\boldsymbol{\beta}}^*, \boldsymbol{\beta}); \boldsymbol{W}] = R[(\hat{\boldsymbol{\beta}}^*, \boldsymbol{\beta}); \boldsymbol{W}],$$

this can be written as

$$\begin{aligned} R[(\hat{\boldsymbol{\beta}}^*, \boldsymbol{\beta}); \boldsymbol{W}] &= nE[(\hat{\boldsymbol{\beta}}^* - \boldsymbol{\beta})'\boldsymbol{W}(\hat{\boldsymbol{\beta}}^* - \boldsymbol{\beta})] \\ &= n\,\mathrm{tr}[\boldsymbol{W}\{E(\hat{\boldsymbol{\beta}}^* - \boldsymbol{\beta})(\hat{\boldsymbol{\beta}}^* - \boldsymbol{\beta})'\}] \\ &= \mathrm{tr}(\boldsymbol{W\Gamma}^*), \end{aligned} \tag{4.10}$$

where $\boldsymbol{\Gamma}^*$ is the covariance matrix of $\hat{\boldsymbol{\beta}}^*$.

The performance of the estimators is generally evaluated by comparing the risk functions with a suitable matrix $\boldsymbol{W}$. Naturally, an estimator with a smaller risk is preferable. The estimator $\hat{\boldsymbol{\beta}}^*$ is called inadmissible if there exists another estimator $\hat{\boldsymbol{\beta}}^0$ such that

$$R(\hat{\boldsymbol{\beta}}^0, \boldsymbol{\beta}) \le R(\hat{\boldsymbol{\beta}}^*, \boldsymbol{\beta}) \quad \text{for all } (\boldsymbol{\beta}, \boldsymbol{W}) \tag{4.11}$$

with strict inequality holding for some $\boldsymbol{\beta}$. If this is the case, then we say that the estimator $\hat{\boldsymbol{\beta}}^0$ dominates $\hat{\boldsymbol{\beta}}^*$. On the other hand, instead of (4.11) holding for every n, we have

$$\lim_{n\to\infty} R(\hat{\boldsymbol{\beta}}^0, \boldsymbol{\beta}) \le \lim_{n\to\infty} R(\hat{\boldsymbol{\beta}}^*, \boldsymbol{\beta}) \quad \text{for all } \boldsymbol{\beta}, \tag{4.12}$$

with strict inequality for some $\boldsymbol{\beta}$, then $\hat{\boldsymbol{\beta}}^*$ is called an asymptotically inadmissible estimator of $\boldsymbol{\beta}$. The expression in (4.11) is usually not easy to prove. An alternative route is to consider the asymptotic distributional risk (ADR).

To begin the process of calculating ADR, we assume the asymptotic cumulative distribution function (ACDF) of $\sqrt{n}(\hat{\boldsymbol{\beta}}^* - \boldsymbol{\beta})/s_e$ exists under the sequence of local alternatives. The ACDF of an estimator $\hat{\boldsymbol{\beta}}^*$ is defined as

$$G(\mathbf{y}) = \lim_{n\to\infty} P[\sqrt{n}(\hat{\boldsymbol{\beta}}^* - \boldsymbol{\beta})/s_e \le \mathbf{y}].$$

The dispersion matrix obtained from ACDF is given by

$$\boldsymbol{\Gamma} = \int\int\cdots\int \mathbf{y}\mathbf{y}' G(\mathbf{y})$$

Now, the ADR is defined as

$$R(\hat{\boldsymbol{\beta}}^*; \boldsymbol{\beta}) = \mathrm{tr}(\boldsymbol{W}\boldsymbol{\Gamma}). \tag{4.13}$$

In passing we would like to remark here that the asymptotic risk may be obtained by replacing $\boldsymbol{\Gamma}$ with the limit of the actual dispersion matrix of $\sqrt{n}(\hat{\boldsymbol{\beta}}^* - \boldsymbol{\beta})$ in the ADR function. However, this may require some extra regularity conditions for consideration.

4.5.1 Bias and ADR Analysis

We obtain the asymptotic distribution of the proposed estimators and the test statistic ψ_n. We use the following theorem.

Theorem 4.1 *Under the regularity conditions, and if $\sigma^2 < \infty$, as $n \to \infty$,*

$$\sqrt{n}\, s_e^{-1}(\hat{\boldsymbol{\beta}}^{UE} - \boldsymbol{\beta}) \xrightarrow{d} N_p(\mathbf{0}, \boldsymbol{C}^{-1}).$$

Bias Analysis

The *asymptotic distributional bias* (ADB) of an estimator $\hat{\boldsymbol{\beta}}^*$ is defined as

$$\mathrm{ADB}(\hat{\boldsymbol{\beta}}^*) = \lim_{n\to\infty} E\left\{n^{\frac{1}{2}}(\hat{\boldsymbol{\beta}}^* - \boldsymbol{\beta})\right\}.$$

Theorem 4.2 *Under the assumed regularity conditions and theorem above, and under* $\{K_n\}$, *the ADB of the estimators are as follows:*

$$ADB(\hat{\boldsymbol{\beta}}_1^{UE}) = \mathbf{0} \tag{4.14}$$

$$ADB(\hat{\boldsymbol{\beta}}_1^{RE}) = -\boldsymbol{C}^{-1}\boldsymbol{H}'(\boldsymbol{H}\boldsymbol{C}^{-1}\boldsymbol{H}')^{-1}\boldsymbol{\omega} \tag{4.15}$$

$$ADB(\hat{\boldsymbol{\beta}}_1^{PE}) = -\boldsymbol{C}^{-1}\boldsymbol{H}'(\boldsymbol{H}\boldsymbol{C}^{-1}\boldsymbol{H}')^{-1}\boldsymbol{\omega} H_{p_2+2}(\chi^2_{p_2,\alpha};\Delta) \tag{4.16}$$

$$ADB(\hat{\boldsymbol{\beta}}_1^{S}) = -(p_2-2)\boldsymbol{C}^{-1}\boldsymbol{H}'(\boldsymbol{H}\boldsymbol{C}^{-1}\boldsymbol{H}')^{-1}\boldsymbol{\omega} E\left\{\chi^{-2}_{p_2};\Delta\right\} \tag{4.17}$$

$$ADB(\hat{\boldsymbol{\beta}}_1^{S+}) = -\boldsymbol{C}^{-1}\boldsymbol{H}'(\boldsymbol{H}\boldsymbol{C}^{-1}\boldsymbol{H}')^{-1}\boldsymbol{\omega}\left[H_{p_2+2}(p_2-2;\Delta) + (p_2-2) \times E\left\{\chi^{-2}_{p_2+2}(\Delta)\right\} + E\left\{\chi^{-2}_{p_2+2}(\Delta) I(\chi^2_{p_2+2}(\Delta) > p_2-2)\right\}\right], \tag{4.18}$$

where

$$E(\chi_p^{-2j}(\Delta)) = \int_0^\infty x^{-2j} d\Phi_p(x;\Delta),$$

and $\Phi_p(x;\Delta)$ is the cdf of a p-variate normal distribution with mean vector $\mathbf{0}$ and covariance matrix Δ. Let $H_{p_2+2}(\cdot\,;\ \Delta)$ be the cdf of a noncentral chi-square distribution with $p_2 + 2$ degrees of freedom and noncentrality parameter Δ.

The bias expressions for all the estimators are not in the scalar form. We therefore take recourse by converting them into quadratic form. Let us define the *asymptotic quadratic distributional bias* (AQDB) of an estimator $\boldsymbol{\beta}^*$ of $\boldsymbol{\beta}$ by

$$AQDB(\boldsymbol{\beta}^*) = [ADB(\boldsymbol{\beta}^*)]'\boldsymbol{\Sigma}^{-1}[ADB(\boldsymbol{\beta}^*)],$$

where $\boldsymbol{\Sigma} = \sigma^2\boldsymbol{C}^{-1}$.

Using the definition and following Ahmed (1997), the asymptotic quadratic distributional biases of the various estimators are presented below.

$$\text{AQDB}(\hat{\boldsymbol{\beta}}_1^{\text{UE}}) = \mathbf{0}, \tag{4.19}$$

$$\text{AQDB}(\hat{\boldsymbol{\beta}}_1^{\text{RE}}) = \Delta \tag{4.20}$$

$$\text{AQDB}(\hat{\boldsymbol{\beta}}_1^{\text{PE}}) = \Delta\left\{H_{p_2+2}(\chi^2_{p_2,\alpha};\Delta)\right\}^2 \tag{4.21}$$

$$\text{AQDB}(\hat{\boldsymbol{\beta}}^{\text{S}}) = (p_2-2)^2\Delta\left[E\left\{\chi^{-2}_{p_2+2}(\Delta)\right\}\right]^2 \tag{4.22}$$

$$\text{ADQB}(\hat{\boldsymbol{\beta}}_1^{\text{S+}}) = \Delta\left[H_{p_2+2}(p_2-2;\Delta) + (p_2-2)E\left\{\chi^{-2}_{p_2+2}(\Delta)\right\} + E\left\{\chi^{-2}_{p_2+2}(\Delta) I(\chi^2_{p_2+2}(\Delta) > p_2-2)\right\}\right]. \tag{4.23}$$

The AQDB of the full model estimator is an unbounded function of Δ. Evidently, the magnitude of its bias will depend on the quantity Δ. The AQDB of the pretest estimator is a function of Δ and size of the pretest α. For fixed value of α, the bias function begins at zero, increases to a certain point, and then decreases gradually to zero. As a function of the size of the test for fixed Δ, the bias function is a decreasing function of $\alpha \in [0, 1)$. The value of AQDB is maximum at $\alpha = 0$ and zero at $\alpha = 1$. Interestingly, the AQDB of the shrinkage estimator starts from zero at $\Delta = 0$, increases to a point, and then decreases toward zero. More importantly, the bias curve of the shrinkage estimator remains below the bias curve of the full model estimator for all values of Δ. On the other hand, the AQDB curve of the positive part estimator remains below the AQDB curve of the shrinkage estimator for all values of Δ.

Risk Analysis

Following Ahmed (2001), we present the risk expressions of the estimators.

Theorem 4.3 *Under the assumed regularity conditions and local alternative* $\{K_n\}$, *the ADR expressions are as follows:*

$$R(\hat{\boldsymbol{\beta}}_1^{\text{UE}}; \boldsymbol{W}) = \sigma^2 \text{tr}(\boldsymbol{W}\boldsymbol{C}^{-1}) \tag{4.24}$$

$$R(\hat{\boldsymbol{\beta}}_1^{\text{RE}}; \boldsymbol{W}) = \sigma^2 \text{tr}(\boldsymbol{W}\boldsymbol{C}^{-1}) - \sigma^2 \text{tr}(\boldsymbol{Q}) + \boldsymbol{\omega}'\boldsymbol{B}^{-1}\boldsymbol{Q}\boldsymbol{\omega} \tag{4.25}$$

$$\begin{aligned} R(\hat{\boldsymbol{\beta}}_1^{\text{PE}}; \boldsymbol{W}) &= \sigma^2 \text{tr}(\boldsymbol{W}\boldsymbol{C}^{-1}) - \sigma^2 \text{tr}(\boldsymbol{Q}) H_{p_2+2}(\chi^2_{p_2,\alpha}; \Delta) \\ &\quad + \boldsymbol{\omega}'\boldsymbol{B}^{-1}\boldsymbol{\omega}\left\{2H_{p_2+2}(\chi^2_{p_2,\alpha}; \Delta) - H_{p_2+4}(\chi^2_{p_2,\alpha}; \Delta)\right\} \end{aligned} \tag{4.26}$$

$$\begin{aligned} R(\hat{\boldsymbol{\beta}}_1^{\text{S}}; \boldsymbol{W}) &= \sigma^2 \text{tr}(\boldsymbol{W}\boldsymbol{C}^{-1}) - (p_2 - 2)\sigma^2 \text{tr}(\boldsymbol{Q}) \left\{2E[\chi^{-2}_{p_2+2}(\Delta)] \right. \\ &\quad \left. - (p_2 - 2)E[\chi^{-4}_{p_2+4}(\Delta)]\right\} \\ &\quad + (p_2 - 2)(p_2 + 2)(\boldsymbol{\omega}'\boldsymbol{B}^{-1}\boldsymbol{Q}\boldsymbol{\omega})E[\chi^{-4}_{p_2+4}(\Delta)] \end{aligned} \tag{4.27}$$

$$\begin{aligned} R(\hat{\boldsymbol{\beta}}_1^{\text{S+}}; \boldsymbol{W}) &= R(\hat{\boldsymbol{\beta}}^{\text{S}_1}; \boldsymbol{W}) + (p_2 - 2)\sigma^2 \text{tr}(\boldsymbol{Q}) \left[E\left\{\chi^{-2}_{p_2+2}(\Delta) I(\chi^2_{p_2+2}(\Delta) \right.\right. \\ &\quad \left.\left. \le p_2 - 2)\right\} - (p_2 - 2)E\left\{\chi^{-4}_{p_2+2}(\Delta) I(\chi^2_{p_2+2}(\Delta) \le p_2 - 2)\right\}\right] \\ &\quad - \sigma^2 \text{tr}(\boldsymbol{Q}) H_{p_2+2}(p_2 - 2; \Delta) + \boldsymbol{\omega}'\boldsymbol{B}^{-1}\boldsymbol{Q}\boldsymbol{\omega}\left\{2H_{p_2+4}(p_2 - 2; \Delta)\right\} \\ &\quad - (p_2 - 2)\boldsymbol{\omega}'\boldsymbol{B}^{-1}\boldsymbol{Q}\boldsymbol{\omega}\left[2E\left\{\chi^{-2}_{p_2+2}(\Delta) I(\chi^2_{p_2+2}(\Delta) \le p_2 - 2)\right\}\right. \\ &\quad - 2E\left\{\chi^{-2}_{p_2+4}(\Delta) I(\chi^2_{p_2+4}(\Delta) \le p_2 - 2)\right\} \\ &\quad \left. + (p_2 - 2)E\left\{\chi^{-4}_{p_2+4}(\Delta) I(\chi^{-4}_{p_2+4}(\Delta) \le p_2 - 2)\right\}\right], \end{aligned} \tag{4.28}$$

where $\boldsymbol{Q} = \boldsymbol{H}\boldsymbol{C}^{-1}\boldsymbol{W}\boldsymbol{C}^{-1}\boldsymbol{H}'\boldsymbol{B}^{-1}, \boldsymbol{B}^{-1} = \boldsymbol{H}'\boldsymbol{C}^{-1}\boldsymbol{H}$.

Ahmed (1997) has examined the risk properties of the listed estimators. It was remarked that none of the full model, submodel, shrinkage, and pretest estimators is inadmissible with respect to any of the others. However, at $\Delta = 0$,

$$R(\hat{\boldsymbol{\beta}}_1^{\mathrm{RE}}; \boldsymbol{W}) < R(\hat{\boldsymbol{\beta}}_1^{\mathrm{PE}}; \boldsymbol{W}) < R(\hat{\boldsymbol{\beta}}_1^{\mathrm{UE}}; \boldsymbol{W}).$$

On the other hand, for $\Delta \geq 0$ and $p_2 \geq 3$,

$$R(\hat{\boldsymbol{\beta}}_1^{\mathrm{S+}}; \boldsymbol{W}) \leq R(\hat{\boldsymbol{\beta}}_1^{\mathrm{S}}; \boldsymbol{W}) \leq R(\hat{\boldsymbol{\beta}}_1^{\mathrm{UE}}; \boldsymbol{W}),$$

with strict inequality holds when $\Delta = 0$. Thus, we conclude that $\hat{\boldsymbol{\beta}}_1^{\mathrm{S+}}$ performs better than $\hat{\boldsymbol{\beta}}_1^{\mathrm{UE}}$ and $\hat{\boldsymbol{\beta}}_1^{\mathrm{S}}$ in the entire parameter space induced by Δ. The gain in risk of all the estimators over the full model estimator is substantial when $\Delta = 0$ or near. Clearly, when Δ moves away from the null hypothesis beyond a certain value, the ADR of the submodel estimator increases and becomes unbounded. This clearly indicates that the performance of the submodel estimator will depend strongly on the reliability of the UPI or AI. The performance of the full model is always steady throughout $\Delta \in [0, \infty)$.

The ADR of the pretest estimator is smaller than the ADR of the full model estimator near the null hypothesis. However, the ADR that keeps on increasing crosses the ADR of the full model estimator, reaches maximum, and then decreases monotonically to the ADR of the full model estimator. Hence a pretest approach controls the magnitude of the ADR. There are points in the parameter space for which the full model estimator is superior to the pretest estimator. We find that the performance of the pretest estimator, which combines data information with UPI, depends heavily on the correctness of this UPI. However, the gain in the ADR can be substantial over the full model estimation strategy when UPI is nearly correct. However, $\hat{\boldsymbol{\beta}}_1^{\mathrm{PE}}$ combines the UPI in a superior way to that of $\tilde{\boldsymbol{\beta}}$, in the sense that the ADR of the pretest estimator is a bounded function of the Δ.

Finally, we compare the ADR performance of shrinkage estimators and the full model estimator. First, we note that under some general conditions

$$R(\hat{\boldsymbol{\beta}}_1^{\mathrm{S+}}; \boldsymbol{W}) \leq R(\hat{\boldsymbol{\beta}}_1^{\mathrm{UE}}; \boldsymbol{W}) \quad \text{for all} \quad \Delta \in [0, \infty),$$

with strict inequality for some Δ. Finally, we may conclude from the ADR relations for these estimators that

$$\frac{R(\hat{\boldsymbol{\beta}}_1^{\mathrm{S+}}; \boldsymbol{W})}{R(\hat{\boldsymbol{\beta}}_1^{\mathrm{UE}}; \boldsymbol{W})} \leq 1, \quad \text{for all} \quad \Delta \in [0, \infty),$$

with strict inequality for some Δ. Therefore, the positive-part shrinkage estimator dominates the shrinkage estimator. Hence, the positive-part shrinkage estimator is also superior to the full model estimator. We observed that the shrinkage estimators combine the information from the full model and the submodel in a superior way, since these estimators perform better than the full model estimator regardless of the correctness of the UPI. However, the gain in ADR over the full model estimator is substantial when the submodel is *nearly* correct. We can also conclude that the proposed positive-part estimator is superior to the shrinkage estimator. However, the important point here is not the improvement in the sense of a lowering of the ADR by using the positive part of the usual Stein-rule estimator. More importantly, the components of the positive-part estimator *have the same sign* as that of components of the full model estimator.

From the above discussion, it can be seen that none of the estimators are inadmissible with respect to each other in the entire parameter space induced by the noncentrality parameter Δ.

4.6 Simulation Studies

This section is divided into three subsections. First, we compare the relative performance of submodel, pretest, and shrinkage estimators to the full model estimator. Next, a real data set is analyzed. Finally, we investigate the relative performance of the penalty estimation.

4.6.1 Full Model, Submodel, Pretest, and Shrinkage Estimation

Monte Carlo simulation experiments have been conducted to examine mean squared error (MSE) performance of the full model, submodel, pretest, and shrinkage estimators. In this study, we simulate the response from the following model:

$$y_i = x_{1i}\beta_1 + x_{2i}\beta_2 + \cdots + x_{pi}\beta_p + \varepsilon_i, \quad i = 1, \ldots, n,$$

where x_{1i} and $x_{2i} \sim N(1, 2)$ independently, and the x_{si} are i.i.d. $N(0, 1)$ for all $s = 3, \ldots, p$ and $i = 1, \ldots, n$. Moreover, ε_i are i.i.d. $N(0, 1)$.

We are interested in investigating the performance of the estimators when a predefined submodel is available. In other words, we are interested in testing the hypothesis $H_0 : \boldsymbol{\beta}_j = \mathbf{0}$, for $j = p_1 + 1, p_1 + 2, \ldots, p_1 + p_2$, with $p = p_1 + p_2$. Accordingly, we partition the regression coefficients as $\boldsymbol{\beta} = (\boldsymbol{\beta}_1, \boldsymbol{\beta}_2) = (\boldsymbol{\beta}_1, \mathbf{0})$. We defined $\Delta = ||\boldsymbol{\beta} - \boldsymbol{\beta}^{(0)}||$, where $\boldsymbol{\beta}^{(0)} = (\boldsymbol{\beta}_1, \mathbf{0})$ and $|| \cdot ||$ is the Euclidean norm. To determine the behavior of the estimators for $\Delta > 0$, further data sets were generated. Various Δ values have been considered.

The simulated MSE performance of an estimator of $\boldsymbol{\beta}_1$ was measured by comparing its MSE with that of the full model estimator as defined below:

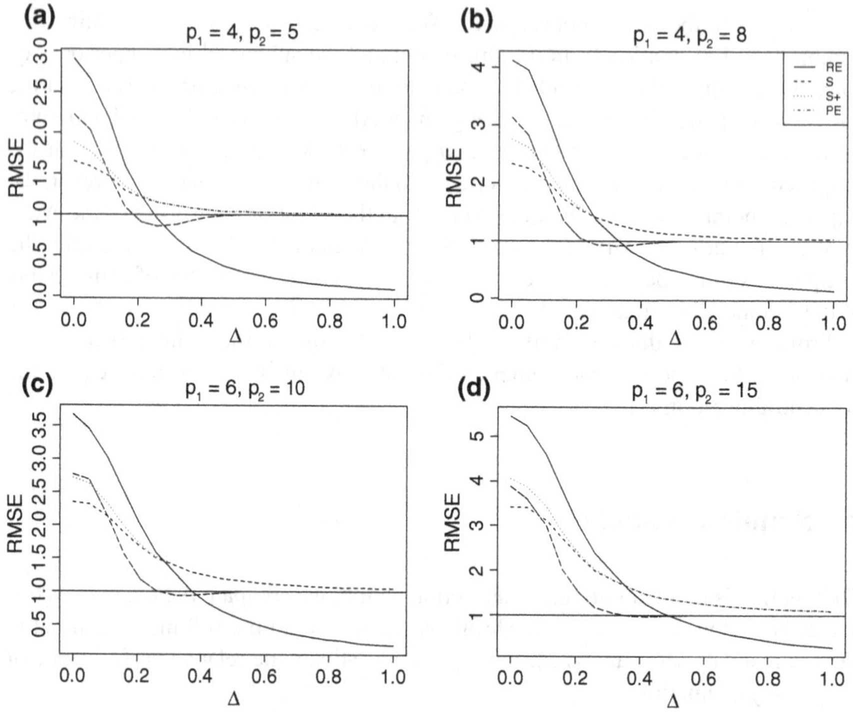

Fig. 4.1 Relative MSE of estimators to full model estimator for $\alpha = 0.05$, $n = 60$, and $(p_1, p_2) = (4, 5), (4, 8), (6, 10)$, and $(6, 15)$

$$\text{RMSE}(\hat{\boldsymbol{\beta}}_1^{\text{UE}} : \hat{\boldsymbol{\beta}}_1^*) = \frac{\text{MSE}(\hat{\boldsymbol{\beta}}_1^{\text{UE}})}{\text{MSE}(\hat{\boldsymbol{\beta}}_1^*)}, \tag{4.29}$$

where $\hat{\boldsymbol{\beta}}_1^*$ is one of the estimators considered in this study. The amount by which an RMSE is larger than unity indicates the degree of superiority of the estimator $\hat{\boldsymbol{\beta}}_1^*$ over the full model estimator.

RMSEs for the proposed estimators are computed for $(p_1, p_2) = (4, 5), (4, 8), (6, 10)$, and $(6, 15)$ for $n = 60$, and plotted in Fig. 4.1. Additionally, Table 4.2 lists the RMSEs for $(p_1, p_2) = (5, 10)$ with $n = 50$. For pretest estimators, we considered $\alpha = 0.05$. Based on Fig. 4.1 and Table 4.2, we draw the following conclusion in the following two scenarios.

Table 4.2 Simulated RMSE of estimators to the full model estimator for $n = 50$, $p_1 = 5$, and $p_2 = 10$

Δ	$\hat{\beta}_1^{\text{RE}}$	$\hat{\beta}_1^{\text{PE}}$	$\hat{\beta}_1^{\text{S}}$	$\hat{\beta}_1^{\text{S+}}$
0.00	4.47	3.37	2.67	3.12
0.05	4.19	2.98	2.53	2.95
0.11	3.65	2.25	2.27	2.55
0.16	3.09	1.73	2.10	2.23
0.21	2.40	1.31	1.84	1.90
0.26	1.92	1.10	1.66	1.69
0.32	1.55	0.97	1.50	1.51
0.37	1.23	0.94	1.40	1.40
0.42	1.01	0.94	1.31	1.31
0.47	0.82	0.96	1.25	1.25
0.53	0.70	0.98	1.21	1.21
0.58	0.59	0.98	1.17	1.17
0.63	0.52	1.00	1.14	1.14
0.68	0.45	1.00	1.13	1.13
0.74	0.39	1.00	1.11	1.11
0.79	0.34	1.00	1.09	1.09
0.84	0.31	1.00	1.09	1.09
0.89	0.27	1.00	1.07	1.07
0.95	0.24	1.00	1.07	1.07
1.00	0.22	1.00	1.06	1.06

Predefined Submodel is True

The simulation study discloses that when the submodel is correctly specified, i.e., $\Delta = 0$, the submodel estimator outperforms all other listed estimators in this study. In other words, at $\Delta = 0$,

$$RMSE(\hat{\boldsymbol{\beta}}_1^{\text{RF}}) > RMSE(\hat{\boldsymbol{\beta}}_1^{\text{PE}}) > RMSE(\hat{\boldsymbol{\beta}}_1^{\text{S+}}) > RMSE(\hat{\boldsymbol{\beta}}_1^{\text{S}}) > 1.$$

Predefined Submodel is Misspecified

In this section, we investigate the performance of the estimators for the general case, i.e., $\Delta > 0$. As $\Delta = 0$ moves away from 0 the RMSE of the submodel converges to 0: in Fig. 4.1 see the sharply decaying curve that goes below the horizontal line at RMSE $= 1$ for $\Delta > 0$. The RMSE of the shrinkage estimators approaches 1 at the slowest rate (for a range of Δ) as we move away from $\Delta = 0$. This indicates that, in the event of an imprecise submodel (i.e., even if $\boldsymbol{\beta}_2 \neq \mathbf{0}$), shrinkage estimators have the largest RMSE among all other estimators for a range of Δ. The shrinkage estimators dominate the full model for all values of Δ. Further, PSE is relatively more efficient than the usual shrinkage estimator, for all values of Δ.

The pretest estimator outshines the shrinkage estimators when Δ is in the neighborhood of zero. Otherwise, for larger values of Δ, the RMSE of the shrinkage estimator is larger than that of the pretest estimator. However, with the increase of Δ, at some point, RMSE of the pretest estimator approaches 1 from below. This phenomenon suggests that neither the pretest nor the submodel estimator is uniformly better than the other when $\Delta > 0$.

To sum up, simulation results suggest that the positive shrinkage estimator maintains its superiority over the full model, and over the submodel and pretest estimators for a wide range of Δ. A positive shrinkage estimator is preferred, as there always remains uncertainty in specifying submodels correctly. Moreover, one cannot go wrong with the shrinkage estimators, even if the assumed submodel is grossly wrong. In such cases, the estimates are as good as or equal to the full model estimates. The findings of the simulation study strongly corroborates the theoretical results presented in Sect. 4.5.

4.6.2 Prostate Data Analysis

Let us recall the prostate example introduced in Sect. 4.4.1. Now, we evaluate the suggested estimation strategy through the prostate data. We compute the average prediction errors and their standard deviations for the full model; and for submodels, pretest and shrinkage estimators for various submodels are shown in Table 4.3. Prediction errors are based on 5 and 10-fold cross-validation. Average and standard errors of the prediction errors are obtained after repeating the process 2,000 times.

In Table 4.3, Submodel (AIC) indicates the submodel estimator when the submodel was selected by AIC. Likewise, Submodel (BIC) and Submodel (BSS) indicate the submodels based on BIC and best subset selection BSS. Similarly, the positive

Table 4.3 Average prediction errors of estimators based on K-fold cross-validation repeated 2,000 times

Estimator	Raw CVE		Bias corrected CVE	
	$K = 5$	$K = 10$	$K = 5$	$K = 10$
Full model	0.556 $_{0.030}$	0.548 $_{0.018}$	0.543 $_{0.026}$	0.542 $_{0.017}$
Submodel (AIC)	0.535 $_{0.023}$	0.529 $_{0.014}$	0.525 $_{0.020}$	0.523 $_{0.013}$
Submodel (BIC)	0.537 $_{0.020}$	0.533 $_{0.012}$	0.529 $_{0.018}$	0.529 $_{0.011}$
Submodel (BSS)	0.582 $_{0.017}$	0.578 $_{0.010}$	0.576 $_{0.015}$	0.576 $_{0.009}$
PS (AIC)	0.554 $_{0.029}$	0.547 $_{0.018}$	0.540 $_{0.025}$	0.541 $_{0.017}$
PS (BIC)	0.546 $_{0.026}$	0.541 $_{0.016}$	0.533 $_{0.023}$	0.535 $_{0.015}$
PS (BSS)	0.549 $_{0.026}$	0.542 $_{0.016}$	0.536 $_{0.023}$	0.536 $_{0.015}$
PE (AIC)	0.536 $_{0.024}$	0.529 $_{0.014}$	0.526 $_{0.021}$	0.525 $_{0.014}$
PE (BIC)	0.538 $_{0.021}$	0.533 $_{0.012}$	0.529 $_{0.019}$	0.529 $_{0.011}$
PE (BSS)	0.599 $_{0.030}$	0.601 $_{0.024}$	0.602 $_{0.036}$	0.605 $_{0.029}$

Numbers in smaller font are the corresponding standard errors shrinkage and pretest estimators are denoted by PS(·) and PE(·), respectively. Looking at the bias corrected cross-validation estimate of the prediction errors, on average, submodel and pretest estimators based on AIC have the smallest prediction errors. This is followed by the pretest and submodel estimators based on BIC. Interestingly, average prediction errors based on the submodel given by BSS are much higher than those obtained from the models based on AIC or BIC. For instance, a submodel based on BSS has an average prediction error of 0.576, and the pretest estimator is of 0.605. For the same submodel, the positive shrinkage estimator has an average prediction error of 0.536, which is much less than the Submodel (BSS) and PE (BSS). Clearly, the positive shrinkage estimator is performing better than the submodel and pretest estimators for this submodel. This is a classic example where the utility of the positive shrinkage estimator is practically realized. Submodel and/or pretest estimation may perform better under correct specification of the submodel (e.g., the models given by AIC and BIC for this data set), whereas the positive shrinkage estimator is less sensitive to the submodel misspecification.

Apparently, in the presence of imprecise auxiliary information, submodel and pretest estimators fail to produce the optimal estimates that reduce average prediction errors. On the other hand, the positive shrinkage estimator maintains a steady risk-superiority under submodel misspecification.

4.6.3 Penalty Estimation

We perform Monte Carlo simulation experiments to examine the relative performance of the penalty estimators to shrinkage estimators and submodel estimators.

We partition the regression coefficients as $\boldsymbol{\beta} = (\boldsymbol{\beta}_1, \boldsymbol{\beta}_2) = (\boldsymbol{\beta}_1, \mathbf{0})$, and consider $\boldsymbol{\beta}_1 = (1, 1, 1, 1, 1)$. Thus, for a fair comparison we are considering the situation that signals can be separated from the noise. Thus, we are dealing with a true sparse model.

The performance of an estimator of $\boldsymbol{\beta}_1$ was measured by calculating its mean squared error (MSE). After calculating the MSEs, we computed RMSE of the estimators to the full model estimator. We include in this study, submodel, shrinkage, and the penalty estimator ($\hat{\boldsymbol{\beta}}^{\text{LASSO}}$, $\hat{\boldsymbol{\beta}}^{\text{aLASSO}}$, and $\hat{\boldsymbol{\beta}}^{\text{SCAD}}$). The notion of RMSE criterion is defined in (4.29).

We present the simulation results for $n = 40, 60$ and $p_2 = 5, 10$ and 15. RMSEs are calculated and are shown in Table 4.4 when the submodel is true. We observe the following:

1. Submodel estimator outperforms all the listed estimators.
2. Positive shrinkage estimator outperforms all the penalty estimators.
3. aLASSO outperforms the LASSO, while SCAD is superior to both LASSO and aLASSO.

However, there are instances when penalty estimators may outperform shrinkage estimators.

Table 4.4 Simulated RMSE for comparing shrinkage and penalty estimators for $p_1 = 5$, $\Delta = 0$

n	p_2	$\hat{\beta}_1^{\text{RE}}$	$\hat{\beta}_1^{\text{S}}$	$\hat{\beta}_1^{\text{S+}}$	$\hat{\beta}_1^{\text{LASSO}}$	$\hat{\beta}_1^{\text{aLASSO}}$	$\hat{\beta}_1^{\text{SCAD}}$
40	5	2.79	1.61	1.84	1.11	1.56	1.63
	10	4.67	2.66	3.16	1.51	2.37	2.64
	15	8.47	4.20	5.06	2.33	4.06	4.37
60	5	2.58	1.59	1.77	1.13	1.59	1.73
	10	4.31	2.75	3.01	1.48	2.41	2.73
	15	7.03	3.72	4.75	2.10	3.57	4.24

Ahmed et al. (2007) was the first to compare shrinkage estimators with an absolute penalty estimator in a partially linear regression setup. Raheem and Ahmed (2012) conducted a detailed study on the comparison of risk performance for shrinkage and some penalty estimators. Similar conclusions were drawn for a partially linear model by Ahmed et al. (2007). In the next chapter we consider the regression parameters problem in a partially linear model.

4.7 Chapter Summary

For a linear regression model, we have considered various estimation strategies for the regression coefficients based on full model, predefined submodel, shrinkage, pretest, and penalty estimation. We derived the asymptotic bias and the risk expressions for the estimators except for the penalty estimators. The relative performance of each estimation procedure is critically assessed.

We showcased two scenarios:

1. When we have prior information about certain covariates, shrinkage estimators are directly obtained by combining the full and submodel estimates.
2. On the other hand, if *a priori* information is not available, the shrinkage estimation takes a two-step approach in obtaining the estimates. In the first step, a set of covariates is selected based on a suitable model selection criterion such as AIC, BIC, or best subset selection. Consequently, the remaining covariates become nuisance, which forms a parametric restriction on the full model, yielding a submodel. In the second step, full and submodel estimates are combined in a way that minimizes the quadratic risk leading to shrinkage estimation.

Shrinkage estimates are obtained using `shrink` (Raheem and Ahmed 2011) R package for shrinkage estimation in linear models. All the calculations have been performed on R statistical software (R Development Core Team 2010).

To illustrate the methods, a real data example has been considered. The suggested estimation strategies are evaluated through application to a real prostate data set. Average prediction errors based on repeated cross-validation estimates of the error rates indicate that shrinkage and submodel estimators have a superior performance compared to penalty estimators when the underlying submodel is correctly specified. This is not unusual, since the submodel estimator dominates all other estimators when the prior information is correct. Since the data considered in this study have been interactively analyzed using various model selection criteria, it is expected that the submodels consist of the best subsets of the available covariates for the respective data sets. Theoretically, this is equivalent to the case where Δ is very close to zero. The real data examples, however, do not tell us how sensitive the prediction errors are under model misspecification. Therefore, we conducted Monte Carlo simulation to study such characteristics for shrinkage and pretest estimators under varying Δ, and different sizes of the nuisance subsets.

In a Monte Carlo study, we numerically computed RMSE for the submodel, shrinkage, and pretest estimators with respect to the full model estimator. Our study re-established the fact that the submodel estimator outperforms the full estimator near the pivot ($\Delta = 0$). However, as we deviate from the pivot ($\Delta > 0$), the MSE of the submodel estimator becomes unbounded. The pretest estimator is a bounded function of Δ. The pretest estimator performs better than the full model estimator for small values of Δ, and its RMSE function approaches from below to merge with the line where RMSE is a unity. On the other hand, RMSE for a positive shrinkage estimator decays at the slowest rate with the increase of Δ, and performs steadily through a wider range of the parameter subspace induced by Δ. The shrinkage estimators dominate the full model estimator.

In penalty estimation, the tuning parameter is estimated using cross-validation. We only do the comparison based on the selected submodel, because the penalty estimators we consider here do not take advantage of the fact that $\boldsymbol{\beta}$ is partitioned into main parameters and nuisance parameters, and thus are at a disadvantage when $\Delta > 0$. Based on our limited simulation study, the shrinkage estimator performs better than the penalty estimators for all the cases considered in the simulation study. In particular, the MSE gain for the shrinkage estimator is more substantial when compared to LASSO than to the SCAD and adaptive LASSO estimators. The penalty estimators are competitive when the number of parameters p_2 in the nuisance parameter vector is small, but the shrinkage estimator with appropriate data-based weights performs best when p_2 is large.

Generally speaking, none of the penalty and shrinkage estimation strategies dominates each other in all situations. Penalty estimation procedure assumed that the signals and noises are well separated. Generally speaking, penalty estimators are not efficient when there are weak signals. This remains as a direction for follow-up research.

References

Ahmed, S., & Basu, A. K. (2000). Least squares, preliminary test and stein-type estimation in general vector ar(p) models. *Statistica neerlandica*, *54*, 47–66.

Ahmed, S. E. (1997). Asymptotic shrinkage estimation: The regression case. *Applied Statistical Science*, *II*, 113–139.

Ahmed, S. E. (2001). Shrinkage estimation of regression coefficients from censored data with multiple observations. In S. Ahmed & N. Reid (Eds.), *Empirical bayes and likelihood inference, lecture notes in statistics* (Vol. 148, pp. 103–120). New York: Springer-Verlag.

Ahmed, S. E., & Chitsaz, S. (2011). Data-based adaptive estimation in an investment model. *Communications in Statistics-Theory and Methods*, *40*, 19–20.

Ahmed, S. E., Doksum, K. A., Hossain, S., & You, J. (2007). Shrinkage, pretest and absolute penalty estimators in partially linear models. *Australian and New Zealand Journal of Statistics*, *49*, 435–454.

Ahmed, S. E., Hossain, S., & Doksum, K. A. (2012). LASSO and shrinkage estimation in Weibull censored regression models. *Journal of Statistical Planning and Inference*, *12*, 1273–1284.

Ahmed, S. E., & Nicol, C. J. (2012). An application of shrinkage estimation to the nonlinear regression model. *Computational Statistics and Data Analysis*, *56*, 3309–3321.

Ahmed, S. E., Raheem, E., and Hossain, M. S. (2010). Absolute penalty estimation. In *International encyclopedia of statistical science*. New York: Springer.

Ahmed, S. E., Saleh, A. K., & Md, E. (1990). Estimation strategies for the intercept vector in a simple linear multivariate normal regression model. *Computational Statistics and Data Analysis*, *10*, 193–206.

Efron, B., Hastie, T., Johnstone, I., & Tibshirani, R. (2004). Least angle regression. *Annals of Statistics*, *32*, 407–499.

Efron, B., & Morris, C. (1972). Empirical bayes on vector observations-an extension of stein's method. *Biometrika*, *59*, 335–347.

Fan, J. (1997). Comments on wavelets in statistics: A review by A. Antoniadis. *Journal of the Italian Statistical Association*, *6*, 131–138.

Fan, J., & Li, R. (2001). Variable selection via nonconcave penalized likelihood and its oracle properties. *Journal of the American Statistical Association*, *96*(456), 1348–1360.

Frank, I. E., & Friedman, J. H. (1993). A statistical view of some chemometrics regression tools. *Technometrics*, *35*, 109–148.

Friedman, J., Hastie, T., & Tibshirani, R. (2010). Regularization paths for generalized linear models via coordinate descent. *Journal of Statistical Software*, *33*(1), 1–22.

Hastie, T., Tibshirani, R., & Friedman, J. (2009). *The elements of statistical learning: Data mining inference and prediction*. New York: Springer.

Khan, B. U., & Ahmed, S. E. (2003). Improved estimation of coefficient vector in a regression model. *Communications in Statistics-Simulation and Computation*, *32*(3), 747–769.

R Development Core Team. (2010). *R: A language and environment for statistical computing*. Vienna, Austria: R Foundation for Statistical Computing. ISBN 3-900051-07-0.

Raheem, S. E., & Ahmed, S. E. (2011). Shrink: An R package for shrinkage estimation in linear regression models. Beta version: R package.

Raheem, S. E., & Ahmed, S. E. (2012). *Shrinkage and absolute penalty estimation in linear models*. (Submitted) WIREs Computational Statistics.

Tibshirani, R. (1996). Regression shrinkage and selection via the lasso. *Journal of the Royal Statistical Society Series B*, 267–288.

Tibshirani, R. J., & Tibshirani, R. (2009). A bias correction for the minimum error rate in cross-validation. *Annals of Applied Statistics*, *3*, 822–829.

Zhang, C. H. (2010). Nearly unbiased variable selection under minimax concave penalty. *Annals of Statistics*, *38*, 894–942.

Zou, H. (2006). The adaptive lasso and its oracle properties. *Journal of the American Statistical Association*, *101*(456), 1418–1429.

Chapter 5
Estimation Strategies in Partially Linear Models

Abstract This chapter showcases the estimation problem in a partially linear regression model, when it is *a priori* suspected that the regression coefficient may be restricted to a subspace. Various asymptotic estimation strategies are presented. These estimators are based on the James–Stein and pretests of significance. We provide natural adaptive estimators that significantly improve upon the full model estimator in the case where some of the predictors may or may not be active for the response of interest. The large sample properties of these estimators are established using the notion of asymptotic distributional risk. Further, we consider the penalty estimation for simultaneous variable selection and regression parameter estimation by using the LASSO, adaptive LASSO, and SCAD strategies. Essentially, we consider two regression models: the full model and the candidate submodel. Generally speaking, Stein-based shrinkage estimation strategy shrinks the full model estimator in the direction of submodel estimator. We appraise the performance of the penalty estimators through Monte Carlo simulation. The properties of all the estimators are compared through simulated mean squared error. Our simulation study reveals that the shrinkage and penalty estimation strategies outperform the full model estimation strategy. Finally, the shrinkage estimators perform better than the penalty estimator when there are many inactive predictors in the model.

Keywords Partially linear models · Pretest and shrinkage estimation · Penalty estimation · Asymptotic bias and risk · Simulation

5.1 Introduction

In this chapter, we consider estimation strategies in a partially linear model where the vector of coefficients $\boldsymbol{\beta}$ in the linear part can be partitioned as $(\boldsymbol{\beta}_1, \boldsymbol{\beta}_2)$ where $\boldsymbol{\beta}_1$ is the coefficient vector for active predictors or main effects (e.g., treatment and/or genetic effects) and $\boldsymbol{\beta}_2$ is a vector for inactive predictors or so-called "nuisance"

DOI: 10.1007/978-3-319-03149-1_5,

effects (e.g., age, location). In this scenario, inference about $\boldsymbol{\beta}_1$ may benefit from moving the benchmark estimate for the candidate full model in the direction of the benchmark estimate without the inactive predictors (Steinian shrinkage), or from deleting the inactive predictors if there is sufficient evidence that these predictors do not provide useful information (pre-testing). We will investigate the large sample properties of Stein-type and pretest semiparametric estimators using quadratic loss function. Interestingly, under general conditions, a Stein-type semiparametric estimator improves on the candidate full model classical semiparametric least square estimator. The relative performance of the estimators is assessed using asymptotic analysis of quadratic risk functions, and it is revealed that the Stein-type estimator outshines the candidate full model estimator in the entire parameter space. However, the pretest estimator dominates the classical estimator only in a small part of the parameter space—a typical characteristic of pretest estimation strategies. We also consider a penalty-type estimator for partially linear models and give a Monte Carlo simulation comparison of shrinkage, pretest, and penalty-type estimators. The comparison shows that the shrinkage method performs better than the penalty-type estimation method when the dimension of the $\boldsymbol{\beta}_2$ parameter space is large.

We consider the partially linear regression model introduced by Engle et al. (1986) to study the effect of weather on electricity demand, in which they assumed that the relationship between temperature and electricity usage was unknown while other related factors such as income and price were parameterized linearly. A partially linear regression model is defined as

$$y_i = \boldsymbol{x}_i'\boldsymbol{\beta} + g(t_i) + \varepsilon_i, i = 1, \ldots, n, \tag{5.1}$$

where y_i's are responses, $\boldsymbol{x}_i = (x_{i1}, \ldots, x_{ip})'$ and $t_i \in [0, 1]$ are design points, $\boldsymbol{\beta} = (\beta_1, \ldots, \beta_p)'$ is an unknown parameter vector, $g(\cdot)$ is an unknown real-valued function defined on [0, 1], and the ε_i's are unobservable random errors.

Model (5.1) has wide applications in sociology, economics, finance, and biometrics. For example, in a clinical trial to compare two treatments, a subject's response will depend on the treatment received and on some covariates (e.g., age). In this case, the experimenter may be unsure of the effect of age on the response, but may want to estimate the treatment differences which are believed to be constant and independent of age (Speckman 1988). A survey of the estimation and application of model (5.1) can be found in the monograph of Härdle et al. (2000). Furthermore, we refer to Wang et al. (2004), Xue et al. (2004), Liang et al. (2004), and Bunea (2004).

Specifying the statistical model is, as always, a critical component in estimation and inference. One typically studies the consequences of some forms of model misspecification. A common type of misspecification in models is caused by including unnecessary predictors in the model, or by leaving necessary (lurking) variables out. The validity of eliminating statistical uncertainty through the specification of a particular parametric formulation depends on information that is generally not available. The aim of this chapter is to analyze some of the issues involved in the estimation of a semiparametric model that may be over-parameterized as discussed in Ahmed et al. (2007) and Hossain et al. (2009). For example, in the data analyzed by Engle et al.

(1986), the electricity demand may be affected by weather, price, income, strikes, and other factors. If we have reason to suspect that a strike has no effect on electricity demand, we may want to decrease the influence of, or delete, this variable.

In many studies the vector of coefficients $\boldsymbol{\beta}$ in (5.1) consists of two components or sub-vectors, one including active and the other including relatively less active or nuisance predictors. For example, if a model is assumed to be sparse then the second sub-vector will be treated as a null vector.

To this end, we partition $(\boldsymbol{\beta}_1', \boldsymbol{\beta}_2')'$, where $\boldsymbol{\beta}_1$ is the coefficient vector of important effects for prediction, for example, treatment effects and genetic effects for the data at hand. In this situation the sub-vector $\boldsymbol{\beta}_2$ may be considered as vector for "nuisance" effects, for example, age and laboratory.

In the above situations, inference about $\boldsymbol{\beta}_1$ will improve by moving the benchmark estimate, obtained from the full model, in the direction of the estimate based on a sub-model. This estimation strategy is known as "Steinian shrinkage". In this framework, the Stein-type estimator combines estimation problems by shrinking a benchmark estimator to a plausible alternative estimator. In the other known approach, one may drop the nuisance variables from the model if there is a statistical evidence that these predictors do not provide useful information, commonly known as pretest estimation.

To formulate the problem, let $\boldsymbol{\beta} = (\boldsymbol{\beta}_1', \boldsymbol{\beta}_2')'$, where $\boldsymbol{\beta}_1$ is the coefficient vector for predicting variables and $\boldsymbol{\beta}_2$ is a vector for those variables which do not play much of a role in overall prediction. To motivate the problem at hand, let us first consider a linear shrinkage strategy. Following the definition of linear shrinkage estimator in Sect. 2.2.1, the linear shrinkage estimator for the partially linear model is defined as

$$\hat{\boldsymbol{\beta}}_1^{\mathrm{LS}} = (1-\pi)\hat{\boldsymbol{\beta}}_1^{\mathrm{RE}} + \pi\hat{\boldsymbol{\beta}}_1^{\mathrm{UE}},$$

where $\hat{\boldsymbol{\beta}}_1^{\mathrm{UE}}$ and $\hat{\boldsymbol{\beta}}_1^{\mathrm{RE}}$ are the estimators of $\boldsymbol{\beta}_1$ for the model with and without the $\boldsymbol{\beta}_2$ sub-vectors, respectively. The constant π is a shrinkage factor and $\pi \in (0, 1)$. For a given data-based $\pi \in (0, 1)$, $\hat{\boldsymbol{\beta}}_1^{\mathrm{LS}}$ improves on the least square estimates $\hat{\boldsymbol{\beta}}_1^{\mathrm{UE}}$ based on the full model, and on the submodel estimate $\hat{\boldsymbol{\beta}}_1^{\mathrm{RE}}$. Burman and Chaudhuri (1992) considered strategies that shrink a nonparametric estimate $\hat{\mu}(\mathbf{x})$ of $\mu(\mathbf{x})$ in the model $Y = \mu(\mathbf{x}) + \varepsilon$ in towards a parametric estimate $g(\hat{\boldsymbol{\beta}}^{\mathrm{UE}}, \mathbf{x})$ of $\mu(\mathbf{x})$. Furthermore, they established conditions under which the suggested estimate asymptotically improves on $\hat{\mu}(\mathbf{x})$ and $g(\hat{\boldsymbol{\beta}}^{\mathrm{UE}}, \mathbf{x})$.

The rest of this chapter is organized as follows. A pretest semiparametric estimator based on the partial kernel method (cf. Speckman 1988) is introduced in Sect. 5.2. Some necessary assumptions are also given in this section. The proposed pretest estimator and shrinkage estimator are presented in Sect. 5.3. The asymptotic properties of the proposed estimators are presented in Sect. 5.4. Results of a simulation study that includes a comparison with a semiparametric extension of the LASSO, aLASSO, and SCAD are given in Sect. 5.5. Finally, the conclusion and some discussions are presented in Sect. 5.6.

5.2 Full and Submodel Estimation Strategies

Let us display again the partial linear model of the form

$$y_i = \boldsymbol{x}_i'\boldsymbol{\beta} + g(t_i) + \varepsilon_i. \tag{5.2}$$

We will assume that $\mathbf{1}_n = (1, \ldots, 1)'$ is not in the space spanned by the column vectors of $\boldsymbol{X} = (\boldsymbol{x}_1, \ldots, \boldsymbol{x}_n)'$. Chen (1988) established that under regularity condition on $g(\cdot)$, this model is identifiable. We further assumed the design points $\boldsymbol{x}_i$ and t_i are fixed for $i = 1, \ldots, n$.

Consider a restriction on the parameters in this model,

$$y_i = \boldsymbol{x}_i'\boldsymbol{\beta} + g(t_i) + \varepsilon_i \quad \text{subject to} \quad \boldsymbol{H\beta} = \boldsymbol{h}, \tag{5.3}$$

where $\boldsymbol{H}$ is an $p_2 \times p$ restriction matrix, and $\boldsymbol{h}$ is an $p_2 \times 1$ vector of constants.

In many applications when the model is sparse then $\boldsymbol{h} = \mathbf{0}$; that is, some of the coefficients are set to zero, and not needed in the initial full model. Let $\boldsymbol{X} = (\boldsymbol{X}_1, \boldsymbol{X}_2)$, where $\boldsymbol{X}_1$ is an $n \times p_1$ submatrix containing the active regression variables of interest, and $\boldsymbol{X}_2$ is a collection $n \times p_2$ submatrix that may or may not be useful in the prediction. Let with $p_1 + p_2 = p$, $p_i \geq 0$ for $i = 1, 2$, then $\boldsymbol{\beta} = (\boldsymbol{\beta}_1', \boldsymbol{\beta}_2')'$ be the vector of parameters, where $\boldsymbol{\beta}_1$ and $\boldsymbol{\beta}_2$ have dimensions p_1 and p_2, respectively.

We are interested in the estimation of $\boldsymbol{\beta}_1$ when the model is sparse. In other words, the sub-vector $\boldsymbol{\beta}_2$ in the full model that is close to $\mathbf{0}$. Thus, we consider a special case, $\boldsymbol{H\beta} = \mathbf{0}$ with $\boldsymbol{H} = (\mathbf{0}, \boldsymbol{I})$, where $\mathbf{0}$ is a $p_2 \times p_1$ matrix of zeroes and $\boldsymbol{I}$ is the identity matrix. Let $\hat{\boldsymbol{\beta}}^{\text{UE}} = (\hat{\boldsymbol{\beta}}_1'^{\text{UE}}, \hat{\boldsymbol{\beta}}_2'^{\text{UE}})'$ be a semiparametric least squares estimator of $\boldsymbol{\beta}$ under full model in (5.1) as defined subsequently. We call $\hat{\boldsymbol{\beta}}_1^{\text{UE}}$ the unrestricted semiparametric least squares estimator of $\boldsymbol{\beta}_1$.

On the other hand, if the model is sparse, that is, $\boldsymbol{\beta}_2 = \mathbf{0}$, then we have the submodel or restricted model which will have the form

$$y_i = x_{i1}\beta_1^{(o)} + \cdots + x_{ip_1}\beta_{p_1}^{(o)} + g^{(o)}(t_i) + \varepsilon_i^{(o)}, \ i = 1, \ldots, n. \tag{5.4}$$

Denoting $\hat{\boldsymbol{\beta}}_1^{\text{RE}}$ as the submodel or restricted semiparametric least squares estimator of $\boldsymbol{\beta}_1$ as defined subsequently. By definition, $\hat{\boldsymbol{\beta}}_1^{\text{RE}}$ performs better than $\hat{\boldsymbol{\beta}}_1^{\text{UE}}$ when the model is actually sparse, that is, the parameter vector $\boldsymbol{\beta}_2$ is close to $\mathbf{0}$. However, when the sub-vector $\boldsymbol{\beta}_2$ moves far away from the pivot $\mathbf{0}$, it will have a drastic effect on estimation and therefore on prediction of $\boldsymbol{\beta}_1$. The submodel estimator, $\hat{\boldsymbol{\beta}}_1^{\text{RE}}$ will be considerably biased, inefficient, and even possibly inconsistent. The estimate $\hat{\boldsymbol{\beta}}_1^{\text{UE}}$ is consistent for departure of $\boldsymbol{\beta}_2$ from $\mathbf{0}$, however may not be efficient, especially when p is large as compared with n. Essentially, we have two extreme estimation strategies, $\hat{\boldsymbol{\beta}}_1^{\text{UE}}$ and $\hat{\boldsymbol{\beta}}_1^{\text{RE}}$ suited best for the partially linear regression models (5.1) and (5.4), respectively. It make sense to consider a compromised strategy between two extreme estimation strategies, so that the compromised strategy behaves reasonably well relative to the full model estimator as well as the submodel estimator.

To this end, we suggest two more estimation strategies for the parameter vector of interest $\boldsymbol{\beta}_1$ of the parametric component in (5.1). The first estimator is the pretest semiparametric least squares estimator, denoted by $\hat{\boldsymbol{\beta}}_1^{\mathrm{PT}}$. To construct a pretest estimator, consider the relevant null hypothesis

$$H_0 : \boldsymbol{\beta}_2 = \mathbf{0}.$$

The pretest estimator is a combination of the full model estimator and submodel estimator via the indicator function $I(T_n < T_{n,\alpha})$, where T_n is an appropriate test function to test the null hypothesis H_0 versus the alternative hypothesis, $H_a : \boldsymbol{\beta}_2 \neq \mathbf{0}$. In addition, $T_{n,\alpha}$ is an α-level critical value using the distribution of T_n. The pretest test estimator selects the full model estimator or submodel estimator based on the outcome of the pretest, that is, whether H_0 is tenable or untenable.

Remark 5.1 It is worth noting that our main objective is to find an efficient estimator of $\boldsymbol{\beta}_1$, keeping in mind that in pretest estimation, deciding against H_a does not necessarily mean we have evidence that $\boldsymbol{\beta}_2$ is a null vector, because we do not have control of the probability of type I error. Alternatively, we hope we may find an efficient estimator of $\boldsymbol{\beta}_1$ by setting $\boldsymbol{\beta}_2 = \mathbf{0}$. Indeed, $T_{n,\alpha}$ is a threshold that determines a hard thresholding rule, and α is a tuning parameter.

The second estimation strategy is based on the Stein-rule estimation, the resulting estimator is known as the shrinkage estimator. From a pretest estimation perspective, the shrinkage estimator $\hat{\boldsymbol{\beta}}_1^{\mathrm{S}}$ can be viewed as a smooth function of the pretest estimator.

In an attempt to estimate the nonparametric component, we confine ourselves to the partial kernel smoothing estimator of $\boldsymbol{\beta}$, which attains the usual parametric convergence rate $n^{-1/2}$ without under-smoothing the nonparametric component $g(\cdot)$, Speckman (1988). Assume that $\{\boldsymbol{x}_i', t_i, y_i; i = 1, \ldots, n\}$ satisfy model (5.1). If $\boldsymbol{\beta}$ is known to be the true parameter, then by $\mathrm{E}(\varepsilon_i) = 0$ we have $g(t_i) = \mathrm{E}(y_i - \boldsymbol{x}_i'\boldsymbol{\beta})$ for $i = 1, \ldots, n$. Hence, a natural nonparametric estimator of $g(\cdot)$ given $\boldsymbol{\beta}$ is

$$\tilde{g}(t, \boldsymbol{\beta}) = \sum_{i=1}^{n} W_{ni}(t)(y_i - \boldsymbol{x}_i'\boldsymbol{\beta}),$$

with the weight functions $W_{ni}(\cdot)$ defined in Assumption 3 below. To estimate $\boldsymbol{\beta}$, we use

$$\hat{\boldsymbol{\beta}}^{\mathrm{UE}} = \operatorname{argmin} SS(\boldsymbol{\beta}) = (\widehat{\boldsymbol{X}}'\widehat{\boldsymbol{X}})^{-1}\widehat{\boldsymbol{X}}'\widehat{\boldsymbol{Y}}, \quad (5.5)$$

with

$$SS(\boldsymbol{\beta}) = \sum_{i=1}^{n} \left(y_i - \boldsymbol{x}_i'\boldsymbol{\beta} - \tilde{g}(t_i, \boldsymbol{\beta})\right)^2 = \sum_{i=1}^{n} (\hat{y}_i - \hat{\boldsymbol{x}}_i'\boldsymbol{\beta})^2,$$

where $\widehat{\boldsymbol{Y}} = (\hat{y}_1, \ldots, \hat{y}_n)'$, $\widehat{\boldsymbol{X}} = (\hat{\boldsymbol{x}}_1, \ldots, \hat{\boldsymbol{x}}_n)'$, $\hat{y}_i = y_i - \sum_{j=1}^{n} W_{nj}(t_i)y_j$ and $\hat{\boldsymbol{x}}_i = \boldsymbol{x}_i - \sum_{j=1}^{n} W_{nj}(t_i)\boldsymbol{x}_j$ for $i = 1, \ldots, n$. The full model or unrestricted estimator $\hat{\boldsymbol{\beta}}_1^{\mathrm{UE}}$ of $\boldsymbol{\beta}_1$ is

$$\hat{\boldsymbol{\beta}}_1^{\mathrm{UE}} = (\widehat{\boldsymbol{X}}_1'\boldsymbol{M}_{\widehat{\boldsymbol{X}}_2}\widehat{\boldsymbol{X}}_1)^{-1}\widehat{\boldsymbol{X}}_1'\boldsymbol{M}_{\widehat{\boldsymbol{X}}_2}\widehat{\boldsymbol{Y}},$$

where $\widehat{\boldsymbol{X}}_1$ is composed of the first p_1 row vectors of $\widehat{\boldsymbol{X}}$, $\widehat{\boldsymbol{X}}_2$ is composed of the last p_2 row vectors of $\widehat{\boldsymbol{X}}$ and $\boldsymbol{M}_{\widehat{\boldsymbol{X}}_2} = \boldsymbol{I} - \widehat{\boldsymbol{X}}_2(\widehat{\boldsymbol{X}}_2'\widehat{\boldsymbol{X}}_2)^{-1}\widehat{\boldsymbol{X}}_2'$.

The submodel or restricted estimator $\hat{\boldsymbol{\beta}}_1^{\mathrm{RE}}$ of $\boldsymbol{\beta}_1$ for model (5.3) is

$$\hat{\boldsymbol{\beta}}_1^{\mathrm{RE}} = (\widehat{\boldsymbol{X}}_1'\widehat{\boldsymbol{X}}_1)^{-1}\widehat{\boldsymbol{X}}_1'\widehat{\boldsymbol{Y}}.$$

We now list the following assumptions required to derive the main results.

Assumption 1 There exist bounded functions $h_s(\cdot)$ over $[0, 1]$, $s = 1, \ldots, p$, such that

$$x_{is} = h_s(t_i) + u_{is}, \; i = 1, \ldots, n, s = 1, \ldots, p, \tag{5.6}$$

where $\boldsymbol{u}_i = (u_{i1}, \ldots, u_{ip})'$ are real vectors satisfying

$$\lim_{n\to\infty} \frac{\sum_{i=1}^{n} u_{ik}u_{ij}}{n} = b_{kj}, \quad \text{for } k = 1, \ldots, p, \; j = 1, \ldots, p, \tag{5.7}$$

and the matrix $\boldsymbol{B} = (b_{kj})$ is nonsingular. Moreover, for any permutation $(j_1, \ldots, j_n)$ of $(1, \ldots, n)$, as $n \to \infty$,

$$\left\| \max_{1\le j\le n} \sum_{i=1}^{n} W_{ni}(t_j)\boldsymbol{u}_i \right\| = o(n^{-\frac{1}{6}}), \tag{5.8}$$

where $||\cdot||$ denotes the Euclidean norm and $W_{ni}(\cdot)$ satisfies Assumption 3.

Assumption 2 The functions $g(\cdot)$ and $h_s(\cdot)$ satisfy the Lipschitz condition of order 1 on $[0, 1]$ for $s = 1, \ldots, p$.

Assumption 3 The probability weight functions $W_{ni}(\cdot)$ satisfy

(i) $\max_{1\le i\le n} \sum_{j=1}^{n} W_{ni}(t_j) = O(1)$,
(ii) $\max_{1\le i,j\le n} W_{ni}(t_j) = O(n^{-2/3})$,
(iii) $\max_{1\le j\le n} \sum_{i=1}^{n} W_{ni}(t_j)I(|t_i - t_j| > c_n) = O(d_n)$, where I is the indicator function, c_n satisfies $\limsup_{n\to\infty} nc_n^3 < \infty$, and d_n satisfies $\limsup_{n\to\infty} nd_n^3 < \infty$.

These assumptions are quite general and can be easily satisfied, see Remarks 5.2–5.3 below.

Remark 5.2 The above u_{ij} behave like zero mean and uncorrelated random variables, and $h_s(t_i)$ is the regression of x_{is} on t_i. Especially, suppose that the design points $(\boldsymbol{x}_i, t_i)$ are i.i.d. random variables, and let $h_s(t_i) = \mathrm{E}(x_{is}|t_i)$ and $u_{is} = x_{is} - h_s(t_i)$ with $\mathrm{E}(\boldsymbol{u}_i\boldsymbol{u}_i') = \boldsymbol{B}$. Then by the law of large numbers, (5.7) holds with probability 1 and (5.8) holds by Lemma 1 in Shi and Lau (2000). Assumptions (5.6) and (5.7) have been used in Gao (1995a,b, 1997), Liang and Härdle (1999), among others, and (5.8) in Shi and Lau (2000).

Remark 5.3 Under regular conditions, the Nadaraya-Watson kernel weights, Priestley and Chao kernel weights, locally linear weights, and Gasser–Müller kernel weights satisfy Assumption 3. For example, if we take the p.d.f. of $U[-1, 1]$ as the kernel function, namely

$$K(t) = I_{[-1,1]}(t)/2,$$

$t_i = i/n$, and the bandwidth is equal to $cn^{-1/3}$, where c is a constant, then the Priestley and Chao kernel weights, which satisfy Assumption 3, are

$$W_{ni}(t) = \frac{1}{2cn^{\frac{2}{3}}} I_{\left(\left|t-\frac{i}{n}\right| \le cn^{-\frac{1}{3}}\right)}(t)\,.$$

5.3 Pretest and Shrinkage Estimation Strategies

This section is devoted to pretest and shrinkage estimation strategies when it is suspected that the full model is sparse.

Pretest Estimation Strategy

As the name says, the pretest estimator is a function or test statistic for testing $H_0 : \boldsymbol{\beta}_2 = \boldsymbol{0}$. We define test statistic T_n as follows:

$$T_n = \frac{n}{\hat{\sigma}_n^2} \hat{\boldsymbol{\beta}}_2' \widehat{\boldsymbol{X}}_2' \boldsymbol{M}_{\widehat{\boldsymbol{X}}_1} \widehat{\boldsymbol{X}}_2 \hat{\boldsymbol{\beta}}_2,$$

where

$$\hat{\sigma}_n^2 = \frac{1}{n}\sum_{i=1}^{n}(y_i - \boldsymbol{x}_i'\hat{\boldsymbol{\beta}} - \hat{g}_n(t_i))^2 = \frac{1}{n}\sum_{i=1}^{n}(\hat{y}_i - \hat{\boldsymbol{x}}_i'\hat{\boldsymbol{\beta}})^2,$$

with $\hat{g}_n(\cdot) = \sum_{i=1}^{n} W_{ni}(\cdot)(y_i - \boldsymbol{x}_i'\hat{\boldsymbol{\beta}})$ and $\boldsymbol{M}_{\widehat{\boldsymbol{X}}_1} = \boldsymbol{I} - \widehat{\boldsymbol{X}}_1(\widehat{\boldsymbol{X}}_1'\widehat{\boldsymbol{X}}_1)^{-1}\widehat{\boldsymbol{X}}_1'$,

Thus, we can choose an α-level critical value $\chi^2_{p_2,\alpha}$ and define $\hat{\boldsymbol{\beta}}_1^{\mathrm{PT}}$ as follows:

$$\hat{\boldsymbol{\beta}}_1^{\mathrm{PT}} = \hat{\boldsymbol{\beta}}_1 - (\hat{\boldsymbol{\beta}}_1 - \hat{\boldsymbol{\beta}}_1^{\mathrm{RE}})I(T_n \leq \chi^2_{p_2,\alpha}).$$

Thus, $\hat{\boldsymbol{\beta}}_1^{\mathrm{PT}}$ has two components and chooses $\hat{\boldsymbol{\beta}}_1^{\mathrm{RE}}$ when the null hypothesis is tenable, otherwise $\hat{\boldsymbol{\beta}}_1$ is selected. Naturally, the dispersion of $\hat{\boldsymbol{\beta}}_1^{\mathrm{PT}}$ is controlled, depending on the value of α, the size of the test. However, the pretest estimation strategy makes extreme choices for either the full model estimator or submodel estimator. As a result, the pretest test procedures are not admissible for many models, even though they may improve on unrestricted procedures, a well-documented fact in the literature. This motivates us to consider another basis for resolving the model-estimator uncertainty. Stein (1956) demonstrated the inadmissibility of the maximum likelihood estimator when estimating a multivariate mean vector under quadratic loss. Sclove et al. (1972) demonstrated the non-optimality of the pretest test estimator in certain multi-parametric situations by making use of Stein-type estimators.

Ahmed (2001) provided expressions for the asymptotic biases and risks of Stein-type estimators in exponential regression models with censored data. Ahmed et al. (2006) studied the asymptotic properties of estimators based on the pretest test and a Stein-rule in a nonparametric model.

The shrinkage semiparametric estimator $\hat{\boldsymbol{\beta}}_1^{\mathrm{S}}$ is defined by

$$\hat{\boldsymbol{\beta}}_1^{\mathrm{S}} = \hat{\boldsymbol{\beta}}_1^{\mathrm{UE}} - (\hat{\boldsymbol{\beta}}_1^{\mathrm{UE}} - \hat{\boldsymbol{\beta}}_1^{\mathrm{RE}})(p_2 - 2)T_n^{-1},\ p_2 \geq 3.$$

Noting that this estimator is in the general form of the Stein-rule family of estimators, where shrinkage of the benchmark estimator is in the direction of the alternative submodel estimator $\hat{\boldsymbol{\beta}}_1^{\mathrm{RE}}$. Interestingly, the shrinkage estimator is pulled toward the submodel estimator when the variance of the least squares estimator is large, and pulled toward the full model least squares estimator when the submodel estimator has high variance, high bias, or is more highly correlated with the least squares estimator. As indicated earlier, now we know that $\hat{\boldsymbol{\beta}}_1^{\mathrm{S}}$ is the smooth version of the pretest estimator, borrowing and extending the language of Donoho and Johnstone (1998), that pretest and shrinkage estimators are based on hard and smooth thresholdings, respectively.

Generally speaking, shrinkage estimators adapt to the magnitude of T_n and tend to $\hat{\boldsymbol{\beta}}_1^{\mathrm{UE}}$ as T_n tends to infinity and to $\hat{\boldsymbol{\beta}}_1^{\mathrm{RE}}$ as $T_n \to p_2 - 2$.

By design, the shrinkage estimator $\hat{\boldsymbol{\beta}}_1^{\mathrm{S}}$ may have a different sign from the full model estimator $\hat{\boldsymbol{\beta}}_1^{\mathrm{UE}}$, caused by over-shrinking. To circumvent this possible over-shrinking issue, we truncate the second term in $\hat{\boldsymbol{\beta}}_1^{\mathrm{S}}$. As a result, this truncation leads to a convex combination of $\hat{\boldsymbol{\beta}}_1^{\mathrm{UE}}$ and $\hat{\boldsymbol{\beta}}_1^{\mathrm{RE}}$. We call the truncated version the positive shrinkage estimator (PSE). This estimator is defined as follows:

$$\hat{\boldsymbol{\beta}}_1^{\mathrm{S+}} = \hat{\boldsymbol{\beta}}_1^{\mathrm{RE}} + \left[1 - \frac{(p_2-2)}{T_n}\right]^+ (\hat{\boldsymbol{\beta}}_1^{\mathrm{UE}} - \hat{\boldsymbol{\beta}}_1^{\mathrm{RE}}),\ p_2 \geq 3,$$

where $z^+ = max(0, z)$. The positive part estimator $\hat{\boldsymbol{\beta}}_1^{\mathrm{S+}}$ is particularly important for controlling over-shrinking.

In passing we remark here that the shrinkage estimation strategy is similar in spirit to the model-averaging strategies, Bayesian or otherwise. For further insight on this remark, we refer to Burnham and Anderson (2002), Hoeting et al. (2002), Hoeting et al. (1999), and Bickel (1984).

We can define another pretest estimator, replacing $\hat{\boldsymbol{\beta}}_1^{\mathrm{UE}}$ by $\hat{\boldsymbol{\beta}}_1^{\mathrm{S}}$ in the pretest estimator. We call this estimator the *improved pretest estimator* (IPT) and is defined by

$$\hat{\boldsymbol{\beta}}_1^{\mathrm{IPT}} = \hat{\boldsymbol{\beta}}_1^{\mathrm{RE}} + \left(1 - (p_2-2)T_n^{-1}\right)(\hat{\boldsymbol{\beta}}_1^{\mathrm{UE}} - \hat{\boldsymbol{\beta}}_1^{\mathrm{RE}})I(T_n > \chi^2_{p_2,\alpha}),\ p_2 \geq 3.$$

This estimator dominates $\hat{\boldsymbol{\beta}}_1^{\mathrm{PT}}$ over the range of parameter values, but we now have the restriction $p_2 \geq 3$. Noting that, if $\chi^2_{p_2,\alpha} \leq (p_2-2)$ then $\hat{\boldsymbol{\beta}}_1^{\mathrm{IPT}}$ behaves like $\hat{\boldsymbol{\beta}}_1^{\mathrm{S+}}$ and, for $\chi^2_{,\alpha}$ outside this range, it behaves like the usual pretest estimator. However, it still continues to perform better than the pretest estimator. See Ahmed (2001) for details.

5.3.1 Penalty Estimators

The penalty estimators are a class of estimators in the penalized least squares family of estimators, see Ahmed et al. (2010).

Several penalty estimators have been proposed in the literature for linear and generalized linear models. In this section, we consider the least absolute shrinkage and selection operation (LASSO), the smoothly clipped absolute deviation method (SCAD), and the adaptive LASSO. By shrinking some regression coefficients to zero, these methods select important variables and estimate the regression model simultaneously.

An important member of the penalized least squares family is the L_1 penalized least squares estimator, and is known as the least absolute shrinkage and selection operator (LASSO).

LASSO

LASSO was proposed by Tibshirani (1996), which performs variable selection and parameter estimation simultaneously. LASSO is closely related with ridge regression. LASSO solutions are similarly defined by replacing the squared penalty $\sum_{j=1}^{p} \beta_j^2$ in the ridge solution (4.5) with the absolute penalty $\sum_{j=1}^{p} |\beta_j|$ in the LASSO,

$$\hat{\boldsymbol{\beta}}^{\text{lasso}} = \underset{\boldsymbol{\beta}}{\operatorname{argmin}} \left\{ \sum_{i=1}^{n} (y_i - \beta_0 - \sum_{j=1}^{p} x_{ij}\beta_j)^2 + \lambda \sum_{j=1}^{p} |\beta_j| \right\}. \tag{5.9}$$

Although the change apparently looks subtle, the absolute penalty term made it impossible to have an analytic solution for the LASSO. Originally, LASSO solutions were obtained via quadratic programming. Later, Efron et al. (2004) proposed Least Angle Regression (LAR), a type of stepwise regression, with which the lasso estimates can be obtained at the same computational cost as that of an ordinary least squares estimation. Furthermore, the LASSO estimator remains numerically feasible for dimensions of p that are much higher than the sample size n.

SCAD

Although the LASSO method does both shrinkage and variable selection due to the nature of the constraint region which often results in several coefficients becoming identically zero, it does not possess oracle properties (Fan and Li 2001). To overcome the inefficiency of traditional variable selection procedures, Fan and Li (2001) proposed SCAD to select variables and estimate the coefficients of variables automatically and simultaneously. This method not only retains the good features of both subset selection and ridge regression, but also produces sparse solutions, ensures continuity of the selected models (for the stability of model selection), and has unbiased estimates for large coefficients. The estimates are obtained as

$$\hat{\boldsymbol{\beta}}^{\text{SCAD}} = \underset{\boldsymbol{\beta}}{\operatorname{argmin}} \left\{ \sum_{i=1}^{n} (y_i - \beta_0 - \sum_{j=1}^{p} x_{ij}\beta_j)^2 + \lambda \sum_{j=1}^{p} p_{\alpha,\lambda}|\beta_j| \right\}.$$

Here $p_{\alpha,\lambda}(\cdot)$ is the smoothly clipped absolute deviation penalty. The solution of SCAD penalty is originally due to Fan (1997). SCAD penalty is a symmetric and a quadratic spline on $[0, \infty)$ with knots at λ and $\alpha\lambda$, whose first-order derivative is given by

$$p_{\alpha,\lambda}(x) = \lambda \left\{ I(|x| \le \lambda) + \frac{(\alpha\lambda - |x|)_+}{(\alpha - 1)\lambda} I(|x| > \lambda) \right\}, \quad x \ge 0. \tag{5.10}$$

Here $\lambda > 0$ and $\alpha > 2$ are the tuning parameters. For $\alpha = \infty$, the expression (5.10) is equivalent to the L_1 penalty.

Adaptive LASSO

Zou (2006) modified the LASSO penalty by using adaptive weights on L_1 penalties on the regression coefficients. Such a modified method was referred to as adaptive LASSO. It has been shown theoretically that the adaptive LASSO estimator is able to identify the true model consistently, and the resulting estimator is as efficient as the oracle.

The adaptive LASSO estimators (aLASSO) $\hat{\beta}^{\text{aLASSO}}$ are obtained by

$$\hat{\boldsymbol{\beta}}^{\text{aLASSO}} = \underset{\boldsymbol{\beta}}{\text{argmin}} \left\{ \sum_{i=1}^{n} (y_i - \beta_0 - \sum_{j=1}^{p} x_{ij}\beta_j)^2 + \lambda \sum_{j=1}^{p} \hat{w}_j |\beta_j| \right\}, \tag{5.11}$$

where the weight function is

$$\hat{w}_j = \frac{1}{|\hat{\beta}_j^*|^{\gamma}}; \quad \gamma > 0,$$

and $\hat{\beta}_j^*$ is a root-n consistent estimator of β. Equation (5.11) is a "convex optimization problem and its global minimizer can be efficiently solved" (Zou 2006).

The penalty-type estimator was first introduced for linear models. We propose the penalty estimators for partially linear models, which is an extension of the penalty estimation methods for linear models. This estimator can be obtained by applying the penalty estimation method to the residuals $(\hat{\boldsymbol{x}}_i, \hat{\boldsymbol{y}}_i)$, $i = 1, 2, \ldots, n$, defined in Sect. 5.2. Note that the output of the penalty estimation resembles shrinkage and pretest methods by both shrinking and deleting coefficients. However, it is different from the pretest and shrinkage procedures because it treats all the covariate coefficients equally. The LASSO does not single out the nuisance covariates for special scrutiny as to their usefulness in estimating main effect coefficients.

Now, we turn our attention to analyzing the performance of suggested estimators. First, we develop the asymptotic properties of the pretest and shrinkage estimators. To provide a meaningful asymptotic analysis, we will consider local Pitman contiguous models where $\boldsymbol{\beta}_2$ depends on n and tends to the zero vector as $n \to \infty$. Such sequences of models have been considered in the estimation context by Bickel (1984) and Claeskens and Hjort (2003), among others. In the following section, we present asymptotic properties of pretest and shrinkage estimators.

5.4 Asymptotic Bias and Risk Analysis

Our main objective is to assess the performance of the full model, submodel, pretest, and shrinkage estimators when $\boldsymbol{\beta}_2$ is close to the null vector. We consider a sequence of local alternatives $\{K_n\}$ given by

$$K_n : \boldsymbol{\beta}_{2(n)} = n^{-\frac{1}{2}}\boldsymbol{\omega}, \ \boldsymbol{\omega} \neq \mathbf{0} \text{ fixed.} \tag{5.12}$$

We study the asymptotic quadratic risks of positive definite matrix (p.d.m.) $\boldsymbol{M}$, by

$$\mathscr{L}(\hat{\boldsymbol{\beta}}_1^*, \boldsymbol{\beta}_1) = n(\hat{\boldsymbol{\beta}}_1^* - \boldsymbol{\beta}_1)'\boldsymbol{M}(\hat{\boldsymbol{\beta}}_1^* - \boldsymbol{\beta}_1),$$

where $\hat{\boldsymbol{\beta}}_1^*$ can be any one of $\hat{\boldsymbol{\beta}}_1^{\text{UE}}$, $\hat{\boldsymbol{\beta}}_1^{\text{RE}}$, $\hat{\boldsymbol{\beta}}_1^{\text{PT}}$, and $\hat{\boldsymbol{\beta}}_1^{\text{S}}$. Now we assume that, for the estimator $\hat{\boldsymbol{\beta}}_1^*$ of $\boldsymbol{\beta}_1$, the asymptotic distribution function of $\hat{\boldsymbol{\beta}}_1^*$ under $\{K_n\}$ exists and is given by

$$F(\boldsymbol{x}) = \lim_{n\to\infty} P\left(\sqrt{n}(\hat{\boldsymbol{\beta}}_1^* - \boldsymbol{\beta}_1) \le \boldsymbol{x}|K_n\right),$$

where $F(\boldsymbol{x})$ is nondegenerate. Then the asymptotic distributional risk (ADR) of $\hat{\boldsymbol{\beta}}_1^*$ is defined as

$$\boldsymbol{R}(\hat{\boldsymbol{\beta}}_1^*, \boldsymbol{M}) = \text{tr}\left(\boldsymbol{M}\int_{\mathscr{R}_{p_1}}\int \boldsymbol{x}\boldsymbol{x}'dF(\boldsymbol{x})\right) = \text{tr}(\boldsymbol{M}\boldsymbol{V}),$$

where $\boldsymbol{V}$ is the dispersion matrix for the distribution $F(\boldsymbol{x})$.

Note that, under non-local (fixed) alternatives, all the estimators are asymptotically equivalent to $\hat{\boldsymbol{\beta}}_1^{\text{UE}}$, while $\hat{\boldsymbol{\beta}}_1^{\text{RE}}$ has an unbounded risk. To obtain the non-degenerate asymptotic distribution F, we consider the local Pitman alternatives (5.12).

First, we present the expression for the asymptotic distributional bias (ADB) of the proposed estimators. The ADB of an estimator $\hat{\boldsymbol{\beta}}_1^*$ is defined as

$$\text{ADB}(\hat{\boldsymbol{\beta}}_1^*) = \lim_{n\to\infty} \text{E}\left(n^{\frac{1}{2}}(\hat{\boldsymbol{\beta}}_1^* - \boldsymbol{\beta}_1)\right).$$

Let $\boldsymbol{B} = \begin{pmatrix} \boldsymbol{B}_{11} & \boldsymbol{B}_{12} \\ \boldsymbol{B}_{21} & \boldsymbol{B}_{22} \end{pmatrix}$ with $\boldsymbol{B}$ defined in Assumption 1, $\Delta = (\boldsymbol{\omega}'\boldsymbol{B}_{22.1}\boldsymbol{\omega})\sigma^{-2}$, $\boldsymbol{B}_{22.1} = \boldsymbol{B}_{22} - \boldsymbol{B}_{21}\boldsymbol{B}_{11}^{-1}\boldsymbol{B}_{12}$, and $H_v(x; \Delta)$ is the cumulative distribution function of the noncentral chi-square distribution with noncentrality parameter Δ and v degrees of freedom. In addition,

$$\text{E}(\chi_v^{-2j}(\Delta)) = \int_0^\infty x^{-2j}dH_v(x; \Delta).$$

then under assumed regularity conditions and $\{K_n\}$, the ADB of the estimators are given in the following theorem.

Theorem 5.1 *The ADB of the estimators are given below.*

$$ADB(\hat{\beta}_1^{\mathrm{UE}}) = \mathbf{0},$$
$$ADB(\hat{\beta}_1^{\mathrm{RE}}) = -\boldsymbol{B}_{11}^{-1}\boldsymbol{B}_{12}\boldsymbol{\omega},$$
$$ADB(\hat{\beta}_1^{PT}) = -\boldsymbol{B}_{11}^{-1}\boldsymbol{B}_{12}\boldsymbol{\omega}H_{p_2+2}(\chi^2_{p_2,\alpha}; (\Delta)),$$
$$ADB(\hat{\beta}_1^{IPT}) = -\boldsymbol{B}_{11}^{-1}\boldsymbol{B}_{12}\boldsymbol{\omega}\Big[H_{(p_2+2)}(p_2-2; \Delta) + E(\chi^{-2}_{p_2+2}(\Delta)I(\chi^2_{p_2+2}(\Delta) > p_2-2))\Big]$$
$$ADB(\hat{\beta}_1^{S}) = -(p_2-2)\boldsymbol{B}_{11}^{-1}\boldsymbol{B}_{12}\boldsymbol{\omega}\mathrm{E}(\chi^{-2}_{p_2+2}(\Delta))$$
$$ADB(\hat{\beta}_1^{S+}) = -\boldsymbol{B}_{11}^{-1}\boldsymbol{B}_{12}\boldsymbol{\omega}\Big[H_{(p_2+2)}(p_2-2, \Delta) + E(\chi^{-2}_{p_2+2}(\Delta)) + (p_2-2)E(\chi^{-2}_{p_2+2}(\Delta))I(\chi^2_{p_2+2}(\Delta) > (p_2-2))\Big].$$

Proof See Ahmed et al. (2007).

For the special case of $\boldsymbol{B}_{12} = \mathbf{0}$, all the estimators are asymptotically unbiased and hence they are equivalent to each other with respect to the ADB measure. Due to this fact, we will confine ourselves to the situation where $\boldsymbol{B}_{12} \neq \mathbf{0}$, and the remaining discussions follow. In this case, the full model estimator is the only asymptotically unbiased estimator of $\boldsymbol{\beta}$, since it is unrelated to the imposed restriction. Noticing that bias expressions of the estimators are in vector form, we convert them into quadratic forms by applying the following simple transformation:

$$(\mathrm{ADB}(\boldsymbol{\omega}))' \boldsymbol{B}_{11.2} (\mathrm{ADB}(\boldsymbol{\omega})),$$

where $\boldsymbol{B}_{11.2} = \boldsymbol{B}_{11} - \boldsymbol{B}_{12}\boldsymbol{B}_{22}^{-1}\boldsymbol{B}_{21}$. Thus,the asymptotic quadratic distributional bias (AQDB) of an estimator $\hat{\boldsymbol{\beta}}_1^*$ of $\boldsymbol{\beta}_1$ by

$$\mathrm{AQDB}(\hat{\boldsymbol{\beta}}_1^*) = (\mathrm{ADB}(\boldsymbol{\delta}))' \boldsymbol{B}_{11.2} (\mathrm{ADB}(\boldsymbol{\delta})),$$

Corollary 5.1 *The AQDB of the estimators are*

$$AQDB(\hat{\beta}_1^{\mathrm{UE}}) = 0,$$
$$AQDB(\hat{\beta}_1^{\mathrm{RE}}) = \gamma,$$
$$AQDB(\hat{\beta}_1^{PT}) = \gamma H^2_{(p_2+2)}(\chi^2_{p_2,\alpha}; \Delta),$$
$$AQDB(\hat{\beta}_1^{IPT}) = \gamma\left[H_{(p_2+2)}(p_2-2, \Delta) + E\left(\chi^{-2}_{p_2+2}(\Delta)I(\chi^2_{p_2+2}(\Delta) > p_2-2\right)\right]^2$$

$$
\begin{aligned}
AQDB(\hat{\boldsymbol{\beta}}_1^{S}) &= (p_2-2)^2\gamma\left(E(\chi_{p_2+2}^{-2}(\Delta))\right)^2\\
AQDB(\hat{\boldsymbol{\beta}}_1^{S+}) &= \gamma\Big[H_{(p_2+2)}(p_2-2,\Delta)+E(\chi_{p_2+2}^{-2}(\Delta)\\
&\quad - E(\chi_{p_2+2}^{-2}(\Delta))I(\chi_{p_2+2}^{2}(\Delta)<(p_2-2))\Big]^2,
\end{aligned}
$$

where $\gamma=\boldsymbol{\omega}'\boldsymbol{B}_{21}\boldsymbol{B}_{11}^{-1}\boldsymbol{B}_{11.2}\boldsymbol{B}_{11}^{-1}\boldsymbol{B}_{12}\boldsymbol{\omega}$.

Evidently, the AQDB of $\hat{\boldsymbol{\beta}}_1^{\mathrm{RE}}$ is an unbounded function of γ. The magnitude of its bias will depend on the quantity of the γ. The quadratic bias of $\hat{\boldsymbol{\beta}}_1^{\mathrm{PT}}$ is a function of γ and α. For fixed α, the bias function begins at zero, increases to a point, then decreases gradually to zero. On the other hand, as a function of α for fixed γ, it is a decreasing function of $\alpha\in[0,1]$, with a maximum value at $\alpha=0$, and is 0 at $\alpha=1$. The bias function $\hat{\boldsymbol{\beta}}_1^{\mathrm{IPT}}$ behaves the same as the bias function of $\hat{\boldsymbol{\beta}}_1^{\mathrm{PT}}$. However, the bias curve of $\hat{\boldsymbol{\beta}}_1^{\mathrm{IPT}}$ remains below the curve of $\hat{\boldsymbol{\beta}}_1^{\mathrm{PT}}$. The AQDB of $\hat{\boldsymbol{\beta}}_1^{\mathrm{S}}$ starts from 0 at $\gamma=0$, and increases to a point, then decreases toward 0. The quadratic bias curve of $\hat{\boldsymbol{\beta}}_1^{\mathrm{S+}}$ remains below the curve of $\hat{\boldsymbol{\beta}}_1^{\mathrm{S}}$.

Under local alternatives and assumed regularity conditions, we obtain the asymptotic dispersion matrices of the estimators by virtue of the following theorem:

Theorem 5.2 *Suppose the assumptions of Theorem 5.1 hold. Then, under* $\{K_n\}$, *the asymptotic covariance matrices of the estimators are*:

$$
\begin{aligned}
\boldsymbol{\Gamma}(\hat{\boldsymbol{\beta}}_1^{\mathrm{UE}}) &= \sigma^2\boldsymbol{B}_{11.2}^{-1},\\
\boldsymbol{\Gamma}(\hat{\boldsymbol{\beta}}_1^{\mathrm{RE}}) &= \sigma^2\boldsymbol{B}_{11}^{-1}+\boldsymbol{B}_{11}^{-1}\boldsymbol{B}_{12}\boldsymbol{\omega}\boldsymbol{\omega}'\boldsymbol{B}_{21}\boldsymbol{B}_{11}^{-1},\\
\boldsymbol{\Gamma}(\hat{\boldsymbol{\beta}}_1^{PT}) &= \sigma^2\left(\boldsymbol{B}_{11.2}^{-1}\left(1-H_{p_2+2}(\chi_{p_2,\alpha}^2;\Delta)\right)+\boldsymbol{B}_{11}^{-1}H_{p_2+2}(\chi_{p_2,\alpha}^2;\Delta)\right)\\
&\quad+\boldsymbol{B}_{11}^{-1}\boldsymbol{B}_{12}\boldsymbol{\omega}\boldsymbol{\omega}'\boldsymbol{B}_{21}\boldsymbol{B}_{11}^{-1}\left(2H_{p_2+2}(\chi_{p_2,\alpha}^2;\Delta)-H_{p_2+4}(\chi_{p_2,\alpha}^2;\Delta)\right),\\
\boldsymbol{\Gamma}(\boldsymbol{\beta}_1^{IPT}) &= \boldsymbol{\Gamma}(\hat{\boldsymbol{\beta}}_1)+(p_2-2)\boldsymbol{B}_{11}^{-1}\boldsymbol{B}_{12}\boldsymbol{B}_{22.1}^{-1}\boldsymbol{B}_{12}\boldsymbol{B}_{11}^{-1}\cdot\\
&\quad\Big[2E(\chi_{p_2+2}^{-2}(\Delta)I(\chi_{p_2+2}^{2}(\Delta)\le\chi_{p_2,\alpha}^2))\\
&\quad-(p_2-2)E(\chi_{p_2+2}^{-4}(\Delta)I(\chi_{p_2+2}^{2}(\Delta)\le\chi_{p_2,\alpha}^2))\Big]\\
&\quad-\sigma^2\boldsymbol{B}_{11}^{-1}\boldsymbol{B}_{12}\boldsymbol{B}_{22.1}^{-1}\boldsymbol{B}_{12}\boldsymbol{B}_{11}^{-1}H_{p_2+2}(\chi_{p_2,\alpha}^2;\Delta)\\
&\quad+\boldsymbol{B}_{11}^{-1}\boldsymbol{B}_{12}\boldsymbol{\omega}\boldsymbol{\omega}'\boldsymbol{B}_{12}\boldsymbol{B}_{11}^{-1}\left[2H_{p_2+2}(\chi_{p_2,\alpha}^2;\Delta)-H_{p_2+4}(\chi_{p_2,\alpha}^2;\Delta)\right]\\
&\quad-(p_2-2)\boldsymbol{B}_{11}^{-1}\boldsymbol{B}_{12}\boldsymbol{\omega}\boldsymbol{\omega}'\boldsymbol{B}_{12}\boldsymbol{B}_{11}^{-1}\Big[2E(\chi_{p_2+2}^{-2}(\Delta)I(\chi_{p_2+2}^{2}(\Delta)\le\chi_{p_2,\alpha}^2))\\
&\quad-2E(\chi_{p_2+4}^{-2}(\Delta)I(\chi_{p_2+4}^{2}(\Delta)\le\chi_{p_2,\alpha}^2))\\
&\quad+(p_2-2)E(\chi_{p_2+4}^{-4}(\Delta)I(\chi_{p_2+4}^{2}(\Delta)\le\chi_{p_2,\alpha}^2))\Big]
\end{aligned}
$$

$$\begin{aligned}
\boldsymbol{\Gamma}(\hat{\boldsymbol{\beta}}_1^S) =& \sigma^2\boldsymbol{B}_{11.2}^{-1} - (p_2-2)\sigma^2\boldsymbol{B}_{11}^{-1}\boldsymbol{B}_{12}\boldsymbol{B}_{22.1}^{-1}\boldsymbol{B}_{21}\boldsymbol{B}_{11}^{-1}\Big[2\mathrm{E}(\chi_{p_2+2}^{-2}(\Delta))\\
&- (p_2-2)\mathrm{E}(\chi_{p_2+2}^{-4}(\Delta))\Big] + (p_2^2-4)\boldsymbol{B}_{11}^{-1}\boldsymbol{B}_{12}\boldsymbol{\omega}\boldsymbol{\omega}'\boldsymbol{B}_{21}\boldsymbol{B}_{11}^{-1}\mathrm{E}(\chi_{p_2+4}^{-4}(\Delta))\\
\boldsymbol{\Gamma}(\hat{\boldsymbol{\beta}}_1^{S+}) =& \boldsymbol{\Gamma}(\hat{\boldsymbol{\beta}}_1^S)\\
&+(p_2-2)\boldsymbol{B}_{21}\boldsymbol{B}_{11}^{-1}\boldsymbol{B}_{11.2}\boldsymbol{B}_{11}^{-1}\boldsymbol{B}_{12}\Big[E(\chi_{p_2+2}^{-2}(\Delta)I(\chi_{p_2+2}^{2}(\Delta)) \le (p_2-2)\\
&-(p_2-2)E(\chi_{p_2+2}^{-4}(\Delta)I(\chi_{p_2+2}^{2}(\Delta)) \le (p_2-2)\Big]\\
&- \boldsymbol{B}_{11}^{-1}\boldsymbol{B}_{12}\boldsymbol{B}_{22.1}^{-1}\boldsymbol{B}_{21}\boldsymbol{B}_{11}^{-1}H_{p_2+2}(p_2-2;\Delta)\\
&+ \boldsymbol{B}_{11}^{-1}\boldsymbol{B}_{12}\boldsymbol{\omega}\boldsymbol{\omega}'\boldsymbol{B}_{21}\boldsymbol{B}_{11}^{-1}[2H_{p_2+2}(p_2-2;\Delta) - H_{p_2+4}(p_2-2;\Delta)]\\
&- (p_2-2)\boldsymbol{B}_{11}^{-1}\boldsymbol{B}_{12}\boldsymbol{\omega}\boldsymbol{\omega}'\boldsymbol{B}_{21}\boldsymbol{B}_{11}^{-1}[2E(\chi_{p_2+2}^{-2}(\Delta)I(\chi_{p_2+2}^{2}(\Delta)) \le (p_2-2)\\
&- 2E(\chi_{p_2+4}^{-2}(\Delta)I(\chi_{p_2+4}^{2}(\Delta)) \le (p_2-2)\\
&+ (p_2-2)E(\chi_{p_2+4}^{-4}(\Delta)I(\chi_{p_2+4}^{2}(\Delta)) \le (p_2-2)].
\end{aligned}$$

Proof See Ahmed et al. (2007), and Hossain et al. (2009).

Using the result of the above theorem, the asymptotic distributional risk (ADR) expressions for the estimators are contained in the following theorem.

Theorem 5.3 *The risk of the estimators are*:

$$\begin{aligned}
R(\hat{\boldsymbol{\beta}}_1^{\mathrm{UE}};\boldsymbol{M}) =& \sigma^2\mathrm{tr}(\boldsymbol{M}\boldsymbol{B}_{11.2}^{-1}),\\
R(\hat{\boldsymbol{\beta}}_1^{\mathrm{RE}};\boldsymbol{M}) =& \sigma^2\mathrm{tr}(\boldsymbol{M}\boldsymbol{B}_{11}^{-1}) + \boldsymbol{\omega}'\boldsymbol{M}\boldsymbol{\omega},\\
R(\hat{\boldsymbol{\beta}}_1^{PT};\boldsymbol{M}) =& \sigma^2\Big(\mathrm{tr}(\boldsymbol{M}\boldsymbol{B}_{11.2}^{-1})\big(1 - H_{p_2+2}(\chi_{p_2,\alpha}^2;\Delta)\big) + \mathrm{tr}(\boldsymbol{M}\boldsymbol{B}_{11}^{-1})H_{p_2+2}(\chi_{p_2,\alpha}^2;\Delta)\Big)\\
&+ \boldsymbol{\omega}'\boldsymbol{M}\boldsymbol{\omega}\Big(2H_{p_2+2}(\chi_{p_2,\alpha}^2;\Delta) - H_{p_2+4}(\chi_{p_2,\alpha}^2;\Delta)\Big),\\
R(\hat{\boldsymbol{\beta}}_1^{S};\boldsymbol{M}) =& \sigma^2\Big(\mathrm{tr}(\boldsymbol{M}\boldsymbol{B}_{11.2}^{-1}) - (p_2-2)\mathrm{tr}(\boldsymbol{M}\boldsymbol{B}_{22.1}^{-1})2\mathrm{E}(\chi_{p_2+2}^{-2}(\Delta))\\
&- (p_2-2)\mathrm{E}(\chi_{p_2+2}^{-4}(\Delta))\Big) + (p_2^2-4)\boldsymbol{\omega}'\boldsymbol{M}\boldsymbol{\omega}\mathrm{E}(\chi_{p_2+4}^{-4}(\Delta)),\\
R(\hat{\boldsymbol{\beta}}_1^{IPT};\boldsymbol{M}) =& R(\hat{\boldsymbol{\beta}}_1^{S}) + (p_2-2)(tr)(\boldsymbol{M}\boldsymbol{\Upsilon})\Big[2E(\chi_{p_2+2}^{-2}(\Delta)I(\chi_{p_2+2}^{2}(\Delta)) \le \chi_{p_2,\alpha}^2)\\
&- (p_2-2)E(\chi_{p_2+2}^{-4}(\Delta)I(\chi_{p_2+2}^{-2}(\Delta) \le \chi_{p_2,\alpha}^2))\Big]\\
&- \sigma^2(tr)(\boldsymbol{M}\boldsymbol{\Upsilon})H_{(p_2+2)}(\chi_{p_2,\alpha}^2;\Delta) + \boldsymbol{\omega}\boldsymbol{M}\boldsymbol{\omega}'\\
&\times\Big[2H_{(p_2+2)}(\chi_{p_2,\alpha}^2;\Delta) - H_{(p_2+4)}(\chi_{p_2,\alpha}^2;\Delta)\Big]\\
&- (p_2-2)\boldsymbol{\omega}\boldsymbol{M}\boldsymbol{\omega}'\{2E(\chi_{p_2+2}^{-2}(\Delta))I(\chi_{p_2+2}^{2}(\Delta) \le \chi_{p_2,\alpha}^2)\\
&- 2E(\chi_{p_2+4}^{-2}(\Delta)I(\chi_{p_2+4}^{2}(\Delta)) \le \chi_{p_2,\alpha}^2)\\
&+ (p_2-2)E(\chi_{p_2+4}^{-4}(\Delta)I(\chi_{p_2+4}^{2}(\Delta)) \le \chi_{p_2,\alpha}^2)\}\\
R(\hat{\boldsymbol{\beta}}_1^{S+};\boldsymbol{M}) =& R(\hat{\boldsymbol{\beta}}_1^{S}) + (p_2-2)(tr)(\boldsymbol{M}\boldsymbol{\Upsilon})\Big[2E(\chi_{p_2+2}^{-2}(\Delta)I(\chi_{p_2+2}^{2}(\Delta)) \le (p_2-2))
\end{aligned}$$

$$
\begin{aligned}
&- (p_2-2)E(\chi^{-4}_{p_2+2}(\Delta))I(\chi^{2}_{p_2+2}(\Delta) \le (p_2-2))\Big] \\
&- (tr)(\boldsymbol{M\Upsilon})H_{(p_2+2)}((p_2-2);\Delta) \\
&+ \boldsymbol{\omega M \omega}'\left[2H_{(p_2+2)}((p_2-2);\Delta) - H_{(p_2+4)}((p_2-2);\Delta)\right] \\
&- (p_2-2)\boldsymbol{\omega M \omega}'\Big[2E(\chi^{-2}_{p_2+2}(\Delta))I(\chi^{2}_{p_2+2}(\Delta) \le (p_2-2)) \\
&- 2E(\chi^{-2}_{p_2+4}(\Delta))I(\chi^{2}_{p_2+4}(\Delta) \le (p_2-2)) \\
&+ (p_2-2)E(\chi^{-4}_{p_2+4}(\Delta))I(\chi^{2}_{p_2+4}(\Delta) \le (p_2-2))\Big].
\end{aligned}
$$

Again we discard the case, in Theorem 5.3 $\boldsymbol{B}_{12} = \boldsymbol{0}$. In this situation $\boldsymbol{B}_{11.2} = \boldsymbol{B}_{11}$. Then the ADR of all estimators reduced to the ADR of $\hat{\boldsymbol{\beta}}_1$. In the remaining discussion, we therefore assume that $\boldsymbol{B}_{12} \neq \boldsymbol{0}$.

If the parametric restriction is true, then it can be verified that

$$
\mathrm{R}(\hat{\boldsymbol{\beta}}_1^{\mathrm{RE}};\boldsymbol{M}) < \mathrm{R}(\hat{\boldsymbol{\beta}}_1^{IPT};\boldsymbol{M}) < \mathrm{R}(\hat{\boldsymbol{\beta}}_1^{\mathrm{PT}};\boldsymbol{M}) < \mathrm{R}(\hat{\boldsymbol{\beta}}_1^{\mathrm{S+}};\boldsymbol{M}) < \mathrm{R}(\hat{\boldsymbol{\beta}}_1^{\mathrm{S}};\boldsymbol{M}) < \mathrm{R}(\hat{\boldsymbol{\beta}}_1^{\mathrm{UE}};\boldsymbol{M}).
$$

First, comparing the shrinkage and full model estimators, we see that $R(\hat{\boldsymbol{\beta}}_1^{S}, \boldsymbol{M})$ satisfies

$$
\begin{aligned}
R(\hat{\boldsymbol{\beta}}_1^{S};\boldsymbol{M}) &= \sigma^2 \mathrm{tr}(\boldsymbol{M}\boldsymbol{B}_{11.2}^{-1}) - (p_2-2)\sigma^2 \mathrm{tr}(\boldsymbol{M}\boldsymbol{B}_{22.1}^{-1}) \\
&\quad \left((p_2-2)\mathrm{E}(\chi^{-4}_{p_2+2}(\Delta)) + \left(1 - \frac{(p_2+2)\sigma^{-2}\boldsymbol{\omega}'\boldsymbol{M}\boldsymbol{\omega}}{2\Delta \mathrm{tr}(\boldsymbol{M}\boldsymbol{B}_{22.1}^{-1})}\right) 2\mathrm{E}(\chi^{-4}_{p_2+4}(\Delta))\right) \\
&\le R(\hat{\boldsymbol{\beta}}_1;\boldsymbol{M}), \text{ for } p_2 \ge 3, \text{ all } \Delta > 0,
\end{aligned}
$$

and for all $\mathbf{M}$ with

$$
\frac{\mathrm{tr}(\boldsymbol{M}\boldsymbol{B}_{22.1}^{-1})}{ch_{\max}(\boldsymbol{M}\boldsymbol{B}_{22.1}^{-1})} \ge \frac{p_2+2}{2},
$$

where $ch_{\max}(.)$ is the maximum characteristic root.

Remark 5.4 The above result was established using the following identity:

$$
\mathrm{E}(\chi^{-2}_{p_2+2}(\Delta)) - (p_2-2)\mathrm{E}(\chi^{-4}_{p_2+2}(\Delta)) = \Delta \mathrm{E}(\chi^{-4}_{p_2+4}(\Delta)),
$$

Thus, for any $\boldsymbol{M} \in \boldsymbol{M}^D$ and $\boldsymbol{\omega}$, $\mathrm{R}(\hat{\boldsymbol{\beta}}_1^{S}, \boldsymbol{M}) \le \mathrm{R}(\hat{\boldsymbol{\beta}}_1^{\mathrm{UE}}, \boldsymbol{M})$ under the local alternative, where

$$
\boldsymbol{M}^D = \left\{\boldsymbol{M} : \frac{\mathrm{tr}(\boldsymbol{M}\boldsymbol{B}_{22.1}^{-1})}{ch_{\max}(\boldsymbol{M}\boldsymbol{B}_{22.1}^{-1})} \ge \frac{p_2+2}{2}\right\}.
$$

As Δ moves away from zero, $\mathrm{R}(\hat{\boldsymbol{\beta}}_1^{\mathrm{RE}}; \boldsymbol{M})$ monotonically increases in Δ and goes to infinity as Δ goes to infinity. The ADR of $\hat{\boldsymbol{\beta}}_1^{\mathrm{UE}}$ remains constant while $\mathrm{R}(\hat{\boldsymbol{\beta}}_1^{\mathrm{PT}}; \boldsymbol{M})$ increases, crossing the line $\mathrm{R}(\hat{\boldsymbol{\beta}}_1^{\mathrm{UE}}; \boldsymbol{M})$ as Δ moves away from a neighborhood of zero.

Moreover, when Δ tends to infinity, the risks of $\hat{\boldsymbol{\beta}}_1^{\mathrm{PT}}$ and $\hat{\boldsymbol{\beta}}_1^{\mathrm{S}}$ approach a common limit; i.e., the risk of $\hat{\boldsymbol{\beta}}_1^{\mathrm{UE}}$. Thus, $\hat{\boldsymbol{\beta}}_1^{\mathrm{PT}}$ and $\hat{\boldsymbol{\beta}}_1^{\mathrm{S}}$ have bounded risks, unlike $\hat{\boldsymbol{\beta}}_1^{\mathrm{RE}}$. For any $\boldsymbol{M} \in \boldsymbol{M}^D$ and all $\boldsymbol{\omega}$, $\mathrm{R}(\hat{\boldsymbol{\beta}}_1^{\mathrm{S+}}) \le \mathrm{R}(\hat{\boldsymbol{\beta}}_1^{\mathrm{S}}) \le \mathrm{R}(\hat{\boldsymbol{\beta}}_1^{\mathrm{UE}})$ under $\{K_n\}$.

In terms of risk, SE dominates UE, and PSE dominates SE. Hence, PSE is also superior to UE.

Both IPT and PT improve on UE at the null hypothesis at the expense of poor performance elsewhere in the parameter space. The magnitude of the risk gain of the pretest estimators over $\hat{\boldsymbol{\beta}}_1^{\mathrm{UE}}$ at the null vector depends on the size α of the test. As α increases, the maximum risk of $\hat{\boldsymbol{\beta}}^{\mathrm{IPT}}$ and $\hat{\boldsymbol{\beta}}^{\mathrm{PT}}$ decreases. If $\chi^2_{p_2,\alpha} \in [0, p_2 - 2]$, then $\hat{\boldsymbol{\beta}}_1^{\mathrm{IPT}}$ can be simply viewed as $\hat{\boldsymbol{\beta}}_1^{\mathrm{S+}}$ and hence $\hat{\boldsymbol{\beta}}_1^{\mathrm{IPT}}$ dominates $\hat{\boldsymbol{\beta}}_1^{\mathrm{UE}}$. On the other hand, $\hat{\boldsymbol{\beta}}^{\mathrm{IPT}}$ behaves like the usual pretest estimator $\hat{\boldsymbol{\beta}}_1^{\mathrm{PT}}$ whenever $\chi^2_{p_2,\alpha} \notin [0, p_2 - 2]$ and hence may no longer be superior to $\hat{\boldsymbol{\beta}}_1^{\mathrm{UE}}$ for all values of $\boldsymbol{\omega}$.

Consider the case $\chi^2_{p_2,\alpha} \in (p_2 - 2, \infty)$: neither $\hat{\boldsymbol{\beta}}_1^{\mathrm{IPT}}$ nor $\hat{\boldsymbol{\beta}}_1^{\mathrm{PT}}$ is superior to $\hat{\boldsymbol{\beta}}_1^{\mathrm{UE}}$ in the entire parameter space. As $\boldsymbol{\omega}$ moves away from the null vector, the value of the risk of $\hat{\boldsymbol{\beta}}_1^{\mathrm{IPT}}$ increases to a maximum after crossing the risk of $\hat{\boldsymbol{\beta}}_1^{\mathrm{UE}}$, then decreases toward it. There are some points in the parameter space where the risk function of $\hat{\boldsymbol{\beta}}_1^{\mathrm{IPT}}$ crosses the risk function of $\hat{\boldsymbol{\beta}}_1^{\mathrm{UE}}$, and hence is subject to the kind of criticism of being absorbed by $\hat{\boldsymbol{\beta}}_1^{\mathrm{PT}}$. Again, $\hat{\boldsymbol{\beta}}_1^{\mathrm{IPT}}$ performs uniformly better than $\hat{\boldsymbol{\beta}}_1^{\mathrm{UE}}$ when $\chi^2_{p_2,\alpha}$ takes the value outside the interval $(p_2 - 2, \infty)$.

Finally, it is important to remark here that the shrinkage estimators for our criteria with $M \in M_D$ outperform the conventional semiparametric least squares estimator in the entire parameter space for $p_2 \ge 3$, while the least square estimator is admissible for $p_2 = 1$ and $p_2 = 2$.

In the following section we present the result of a simulation study to illustrate the properties of the theoretical results for moderate and large sample sizes. We investigate the relative performance of the listed estimators including penalty estimators using the simulated data.

5.5 Simulation Study

In this section, we use simulated data to investigate the mean squared error (MSE) performance of all the suggested estimators.

The Monte carlo experiment consists of different combinations of sample sizes, i.e., $n = 30, 50, 80$, and 100.

In this study we simulate the response from the following partial linear model:

$$y_i = x_{1i}\beta_1 + x_{2i}\beta_2 + \cdots + x_{pi}\beta_p + g(t_i) + \varepsilon_i, \quad i = 1, \ldots, n,$$

where the ε_i are i.i.d standard normal, $t_i = (i - 0.5)/n$, $x_{si} = (\zeta_{si}^{(1)})^2 + \zeta_i^{(1)}$ with $\zeta_{si}^{(1)}$ i.i.d. $\sim N(0, 1)$ and $\zeta_i^{(2)}$ i.i.d. $\sim N(0, 1)$ for all $s = 1, \ldots, p$, and $i = 1, \ldots, n$.

We consider the the parametric restriction in the form of null hypothesis, H_0 : $\beta_j = 0$, for $j = p_1 + 1, \ldots, p$ with $p = p_1 + p_2$. We set the regression coefficients $\boldsymbol{\beta} = (\boldsymbol{\beta}_1, \boldsymbol{\beta}_2) = (\boldsymbol{\beta}_1, \mathbf{0})$ with $\boldsymbol{\beta}_1 = (1.5, 3, 2)$, and the nonlinear function $g(t) = \sin(4\pi t)$ to generate response y_i. These are fixed for each realization.

In our simulation study, we use

$$W_{ni}(t_j) = \frac{1}{nh_n} K\left(\frac{t_i - t_j}{h_n}\right) = \frac{1}{nh_n}\frac{1}{\sqrt{2\pi}} e^{-\frac{(t_i - t_j)^2}{2h_n^2}},$$

namely Priestley and Chao's weight with a Gaussian kernel.

Also, in our simulation experiment, we impliment the cross-validation method (Bowman and Azzalini 1997) to select the optimal bandwidth h_n, which minimizes the following function:

$$g(h_n) = \frac{1}{n}\sum_{i=1}^{n}(\hat{y}^{-i} - \hat{x}_1^{-i}\hat{\beta}_{1n}^{-i} - \hat{x}_2^{-i}\hat{\beta}_{2n}^{-i} - \hat{x}_3^{-i}\hat{\beta}_{3n}^{-i} - \hat{x}_4^{-i}\hat{\beta}_{4n}^{-i} - \cdots - \hat{x}_p^{-i}\hat{\beta}_{pn}^{-i})^2,$$

where $(\hat{\beta}_{1n}^{-i}, \hat{\beta}_{2n}^{-i}, \hat{\beta}_{3n}^{-i}, \hat{\beta}_{4n}^{-i})' = (\hat{\boldsymbol{X}}'^{-i}\hat{\boldsymbol{X}}^{-i})^{-1}\hat{\boldsymbol{X}}'^{-i}\hat{\boldsymbol{y}}^{-i}$, $\hat{\boldsymbol{X}}^{-i} = \left(\hat{x}_{jk}^{-i}\right)'$, $1 \le k \le n$, $1 \le j \le p$, $\hat{\boldsymbol{y}}^{-i} = (\hat{y}_1^{-i}, \ldots, \hat{y}_n^{-i})$, $\hat{x}_{sk}^{-i} = x_{sk} - \sum_{j \neq i}^{n} W_{nj}(t_i)x_{sj}$, $\hat{y}_k^{-i} = y_k - \sum_{j \neq i}^{n} W_{nj}(t_i)y_j$. Here $\hat{\boldsymbol{y}}^{-i}$ is the predicted value of $\boldsymbol{y} = (y_1, y_2, \ldots y_n)$ at $\boldsymbol{x}_i = (x_{1i}, x_{2i}, \ldots, x_{pi})$ with y_i and $\boldsymbol{x}_i$ left out of the estimation of the β's.

For comparison purposes, we define the parameter $\Delta^\star = ||\boldsymbol{\beta} - \boldsymbol{\beta}^{(0)}||^2$, where $\boldsymbol{\beta}^{(0)} = (\boldsymbol{\beta}_1, \boldsymbol{0})'$ and $|| \cdot ||$ is the Euclidian norm. More importantly, we investigate the characteristic of the suggested estimators for $\Delta^\star > 0$. To do so, further samples were generated such that $\Delta^\star > 0$.

We numerically assessed the performance of an estimator of $\boldsymbol{\beta}_1$ based on the MSE criterion. However, for relative performance, we have computed the relative MSE of all the estimators relative to $\hat{\boldsymbol{\beta}}_1^{\mathrm{UE}}$.

The *relative mean squared error* (RMSE) of an estimator $\hat{\boldsymbol{\beta}}_1^{\diamond}$ to the full model least square estimator $\hat{\boldsymbol{\beta}}_1^{\mathrm{UE}}$ is defined as follows:

$$\mathrm{RMSE}(\hat{\boldsymbol{\beta}}_1^{\mathrm{UE}} : \hat{\boldsymbol{\beta}}_1^{\diamond}) = \frac{\mathrm{MSE}(\hat{\boldsymbol{\beta}}_1^{\mathrm{UE}})}{\mathrm{MSE}(\hat{\boldsymbol{\beta}}_1^{\diamond})}.$$

Table 5.1 RMSE of the estimators to the full model estimator for $n = 60$, $p_2 = 6$

$\Delta^\star$	$\hat{\beta}_1^{RE}$	$\hat{\beta}_1^{IPT}$	$\hat{\beta}_1^{S+}$
0.0	2.69	2.48	1.88
0.2	1.63	1.24	1.36
0.4	0.74	1.03	1.20
0.6	0.38	1.02	1.06
0.8	0.26	1.03	1.03
1.0	0.11	1.02	1.02

Evidently, the amount by which an RMSE is larger than one indicates the degree of superiority of the estimator $\hat{\boldsymbol{\beta}}_1^{\diamond}$ over $\hat{\boldsymbol{\beta}}_1^{\text{UE}}$.

Our methods were applied to several simulated data sets. We report the result in Tables 5.1, 5.2 and Fig. 5.1. Now, we provide the analysis based on simulated data.

Undoubtedly, from Table 5.1 when $\Delta^\star$ is near the origin, the submodel estimator outperforms all the suggested estimators in the class. On the contrary, when $\Delta^\star$ is larger than zero, the estimated MSE of $\hat{\boldsymbol{\beta}}_1^{\text{RE}}$ increases and becomes unbounded, whereas the estimated RMSEs of all other estimators remain bounded and approach one.

Clearly, the departure from the restriction is fatal to $\hat{\boldsymbol{\beta}}_1^{\text{RE}}$, but it has a much smaller impact on the shrinkage and pretest estimators. The finding is consistent with the asymptotic theory.

The pretest estimator behaves well near the null hypothesis. Our simulation study clearly indicates that the performance heavily depends on how close $\Delta^\star$ is to zero. As a result, the pretest estimator is relatively less efficient than the full model least square estimator $\hat{\boldsymbol{\beta}}_1^{\text{UE}}$ for large values of $\Delta^\star$. More importantly, unlike submodel estimator, the MSE of the pretest estimator is a bounded function of $\Delta^\star$. The performance of the improved pretest estimator is similar to that of the pretest estimator. However, it is more efficient than the usual pretest estimator.

The shrinkage estimator has remarkable MSE performance. At $\Delta^\star = 0$, it is highly efficient than $\hat{\boldsymbol{\beta}}_1^{\text{UE}}$. On the other hand, as $\Delta^\star$ increases, the RMSE of the shrinkage estimator decreases and converges to one irrespective of p_1, p_2, and n. Figure 5.1 shows these relative efficiencies of the estimators when the restricted parameter space is correct and incorrect ($\Delta \geq 0$). Furthermore, Fig. 5.1 shows that the shrinkage estimator works better in cases with large p_2. The positive part estimator MSE function behaves the same way as that of the shrinkage estimator that is dominating the full model estimator in the entire parameter space induced by $\Delta^\star$. More importantly, it is a superior strategy to the usual shrinkage estimation strategy.

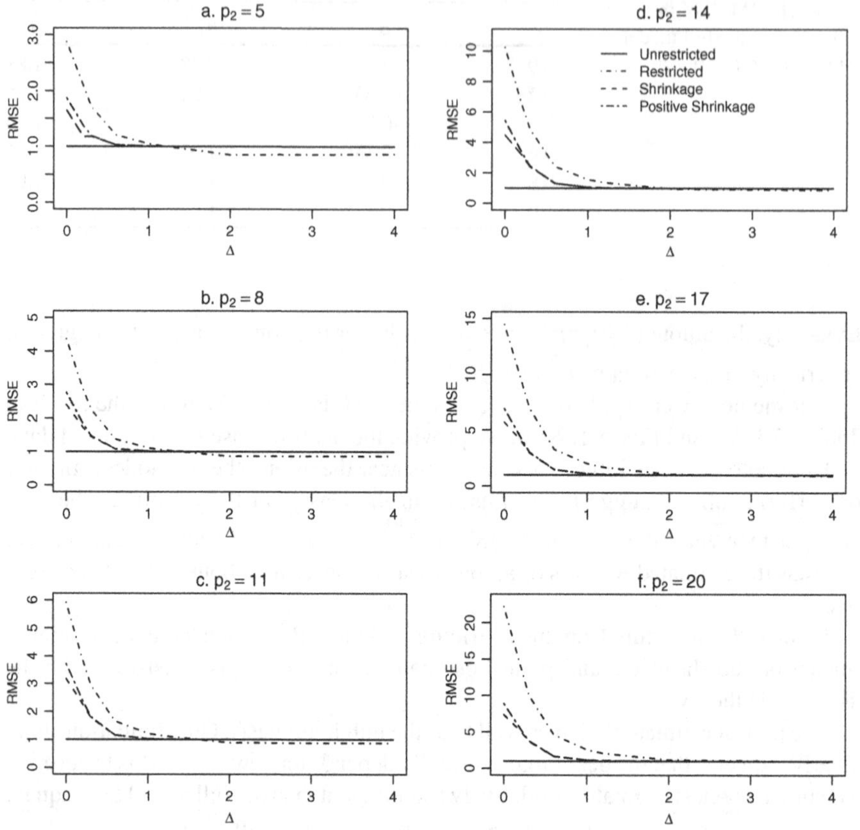

Fig. 5.1 RMSE of the listed estimators to the full model estimator for $p_1 = 4$, $p_2 = 5, 8, 11, 14, 17, 20$, and $n = 60$

5.5.1 Penalty Estimation Strategies

We showcase the result of simulation study pertaining to penalty estimators; and compared with its competitors. The results based on the simulated data are presented in Table 5.2 for the simulated relative efficiencies of RE, SE, PSE, LASSO, adaptive LASSO, and SCAD with respect to the full model estimator when the restricted parameter space $\boldsymbol{\beta}_2 = \mathbf{0}$ is correct ($\Delta^\star = 0$). Table 5.2 reveals that restricted, shrinkage, and positive shrinkage estimators outperform all other estimators in terms of risk. Among the shrinkage and penalty estimators, positive shrinkage estimators maintain superiority over LASSO, aLASSO, and SCAD estimators when p_2 is relatively large.

All computations were conducted using the *R* statistical system (R Development Core Team (2010)).

Table 5.2 RMSE of the proposed estimators to the full model estimator assuming submodel is true

Method	$p_2 = 5$	$p_2 = 8$	$p_2 = 11$	$p_2 = 14$	$p_2 = 17$	$p_2 = 20$
RE	2.88	4.37	5.91	10.23	14.46	22.18
SE	1.64	2.40	3.17	4.46	5.70	7.34
PSE	1.87	2.75	3.71	5.41	6.91	8.85
LASSO	1.43	1.91	2.35	3.64	5.27	6.09
ALASSO	1.51	1.98	2.43	3.88	5.47	6.68
SCAD	1.54	2.16	2.99	4.85	6.19	7.19

5.6 Chapter Summary

In this chapter, in the context of a partially linear regression model with potentially irrelevant nuisance variables, we study the relative performance of full model, submodel, shrinkage, pretest, and penalty estimators.

Using the notion of asymptotic distributional risk, we critically and judiciously examined the risk performance of the estimators. Further, we appraise the risk properties of penalty estimators with other estimators using Monte Carlo experiments.

We conclude, both analytically and numerically, that the submodel estimator and the pretest estimator dominate the full model least square estimators at and near the null hypothesis. The penalty estimators are relatively more efficient when p_2 in the nuisance parameter vector $\boldsymbol{\beta}_2$ has low dimension. Interestingly, the adaptive shrinkage estimators perform best when p_2 is large. Evidently, the shrinkage estimators outshine the full model least squares estimator in the entire parameter space induced by the restriction. In contrast, the performance of the submodel estimator heavily depends on the nuisance effect. Not only that, but the risk of this estimator tends to become unbounded when such submodel does not hold. The risk of the pretest estimators is smaller than the risk of the full model estimator $\hat{\boldsymbol{\beta}}_1^{\mathrm{UE}}$ at or near the null hypothesis. On the other hand, as the hypothesis error grows, the magnitude of the risk of the pretest estimators increases, crosses the risk of the full model estimator,reaches to a maximum value, then decreases monotonically toward the risk of $\hat{\boldsymbol{\beta}}_1^{\mathrm{UE}}$.

The shrinkage and pretest estimation strategies can be extended in various directions to more complex problems. Research on the statistical implications of proposed and related estimators is ongoing. An important problem will be to study the asymptotic properties of pretest and shrinkage estimators when p increases with n and/or $n < p$. It is worth mentioning that this is one of the two areas Bradley Efron predicted for the early twenty-first century (RSS News, January 1995). Shrinkage and likelihood-based methods continue to be extremely useful tools for efficient estimation.

References

Ahmed, S. E. (2001). Shrinkage estimation of regression coefficients from censored data with multiple observations. In S. Ahmed & N. Reid (Eds.), *Empirical Bayes and Likelihood Inference, Lecture Notes in Statistics* (Vol. 148, pp. 103–120). New York: Springer-Verlag.

Ahmed, S. E., Doksum, K. A., Hossain, S., & You, J. (2007). Shrinkage, pretest and absolute penalty estimators in partially linear models. *Australian and New Zealand Journal of Statistics, 49*, 435–454.

Ahmed, S. E., Hussein, A. A., & Sen, P. K. (2006). Risk comparison of some shrinkage m-estimators in linear models. *Nonparametric Statistics, 18*(4–6), 401–415.

Ahmed, S. E., Raheem, E., & Hossain, M. S. (2010). *International Encyclopedia of Statistical Science, chapter Absolute Penalty Estimation*. New York: Springer.

Bickel, P. (1984). Parametric robustness: small biases can be worthwhile. *Annals of Statistics, 12*, 864–879.

Bowman, A., & Azzalini, A. (1997). *Applied Smoothing Techniques for Data Analysis: The Kernel Approach with S-Plus Illustrations*. Oxford: Oxford University Press.

Bunea, F. (2004). Consistent covariate selection and post model selection inference in semiparametric regression. *Annals of Statistics, 32*, 898–927.

Burman, P. & Chaudhuri, P. (1992). *A Hybrid Approach to Parametric and Nonparametric Regression*. Technical Report No. 243, Division of Statistics. USA: University of California-Davis.

Burnham, K., & Anderson, D. (2002). *Model selection and multimodel inference*. New York: Springer-Verlag.

Chen, H. (1988). Convergence rates for parametric components in a partially linear model. *Annals of Statistics, 16*, 136–147.

Claeskens, G., & Hjort, N. (2003). The focused information criterion. *Journal of the American Statistical Association, 98*, 900–916.

Donoho, D., & Johnstone, I. (1998). Minimax estimation via wavelet shrinkage. *Annals of Statistics, 26*, 879–921.

Efron, B., Hastie, T., Johnstone, I., & Tibshirani, R. (2004). Least angle regression. *Annals of Statistics, 32*, 407–499.

Engle, R. F., Granger, W. J., Rice, J., & Weiss, A. (1986). Semiparametric estimates of the relation between weather and electricity sales. *Journal of the American Statistical Association, 80*, 310–319.

Fan, J. (1997). Comments on wavelets in statistics: A review by a. antoniadis. *Journal of the Italian Statistical Association, 6*, 131–138.

Fan, J., & Li, R. (2001). Variable selection via nonconcave penalized likelihood and its oracle properties. *Journal of the American Statistical Association, 96*(456), 1348–1360.

Gao, J. T. (1995a). Asymptotic theory for partially linear models. *Communications in Statistics-Theory and Methods, A, 24*(8), 1985–2009.

Gao, J. T. (1995b). The laws of the iterated logarithm of some estimates in partially linear models. *Statistics and Probability Letters, 25*, 153–162.

Gao, J. T. (1997). Adaptive parametric test in a semiparametric regression model. *Communications in Statistics-Theory and Methods, A26*, 787–800.

Härdle, W., Liang, H., & Gao, J. (2000). *Partially Linear Models*. Heidelber: Physica-Verlag.

Hoeting, J., Madigan, D., Rafferty, A., & Volinsky, C. (1999). *Bayesian model averaging: A tutorial. Statistical Science,14*, 382–401.

Hoeting, J., Rafferty, A., & Madigan, D. (2002). Bayesian variable and transformation selection in linear regression averaging: A tutorial. *Journal of Computational and Graphics Statistics, 11*, 485–507.

Hossain, S., Doksum, K. J., & Ahmed, S. E. (2009). Positive shrinkage, improved pretest and absolute penalty estimators in partially linear models. *Linear Algebra and its Applications, 430*, 2749–2761.

Liang, H., & Härdle, W. (1999). Large sample theory of the estimation of the error distribution for a semiparametric model. *Communications in Statistics-Theory and Methods*, *A28*, 2025–2037.
Liang, H., Wong, S., Robins, J., & Carroll, R. (2004). Estimation in partially linear models with missing covariates. *Journal of the American Statistical Association*, *99*, 357–367.
R Development Core Team. (2010). *R: A Language and Environment for Statistical Computing*. Austria: R Foundation for Statistical Computing. ISBN 3-900051-07-0.
Sclove, S. L., Morris, C., & Radhakrishnan, R. (1972). Non optimality of preliminary test estimation for the multinormal mean. *The Annals of Mathematical Statistics*, *43*, 1481–1490.
Shi, J., & Lau, T. (2000). Empirical likelihood for partially linear models. *Journal of Multivariate Analysis*, *72*, 132–149.
Speckman, P. (1988). Kernel smoothing in partial linear models. *Journal of the Royal Statistical Society Series B*, *50*, 413–437.
Stein, C. (1956). Inadmissibility of the usual estimator for the mean of a multivariate normal distribution. In: *Proceeding of the fourth Berkeley symposium on mathematical statistics an probability*, University of California Press, Berkeley, CA vol 1, 197–206.
Tibshirani, R. (1996). Regression shrinkage and selection via the lasso. *Journal of the Royal Statistical Society Series B*, *58*, 267–288.
Wang, Q., Linton, O., & Härdle, W. (2004). Semiparametric regression analysis with missing response at random. *Journal of the American Statistical Association*, *99*, 334–345.
Xue, H., Lam, K. F., & Gouying, L. (2004). Sieve maximum likelihood estimator for semiparametric regression models with current status data. *Journal of American the Statistical Association*, *99*, 346–356.
Zou, H. (2006). The adaptive lasso and its oracle properties. *Journal of the American Statistical Association*, *101*(456), 1418–1429.

Chapter 6
Estimation Strategies in Poisson Regression Models

Abstract We consider the application of shrinkage and penalty estimation for a Poisson regression model. We present a large sample theory for the full model, submodel, and shrinkage estimators in terms of their respective asymptotic bias and risk. Generally speaking, shrinkage estimators are more efficient than the full model estimator. Nowadays, variable selection is of fundamental importance for modeling and data analysis. A number of variable selection approaches have been proposed in the literature. On the other hand, absolute penalty estimation strategy is useful for simultaneous variable selection and estimation. For this purpose, we consider three penalty estimators, namely, LASSO, adaptive LASSO, and SCAD. We assess the relative performance of the penalty estimators with the shrinkage estimators using Monte Carlo simulation. The relative performance of each estimation strategy is given in terms of a simulated mean squared error. The simulation results reveal that shrinkage method is an effective consistent model selection technique and is comparable to the LASSO, adaptive LASSO, and SCAD when the model is sparse and number predictors in the model is weak. Finally, the listed estimation strategies are appraised through the application to two real data sets.

Keywords Poisson regression models · Pretest and shrinkage estimation · Penalty estimation · Asymptotic bias and risk · Simulation

6.1 Introduction

It has been known that shrinkage estimation strategies produce estimators which are far superior in terms of risk to the maximum likelihood estimator over the entire parameter space. However, up until relatively recently, these estimators have only been used to a limited extent in applications, in part, owing to the computational burden for the purposes of statistical inference. Now with rapid advancement in computing capability, and clear advantages to the use of prior information in certain

DOI: 10.1007/978-3-319-03149-1_6, © The Author(s) 2014

applications, this picture is changing. For instance, numerous cases of shrinkage estimation have appeared in applications involving the real estate market, where appraisers' expert knowledge can be very informative, or in housing pricing models where real estate experts' knowledge and expertise often yield precise information regarding certain parameter values.

More importantly, shrinkage estimators are relatively more efficient than classical estimators based on full model and submodel, respectively. While it is true that submodel estimators can offer a substantial risk gain over full model estimators, there is still a concern that submodel estimators are less desirable to use when *uncertain prior information* (UPI) or *auxiliary information* (AI) is incorrect. The advantage of the shrinkage approach is, therefore, that UPI or AI is incorporated into estimation to the extent that it appears to be true, given sample information. The use of shrinkage estimators is an attractive and effective trade-off in the context of numerous applications.

We consider the application of shrinkage and penalty estimation to a Poisson regression model. The Poisson regression model is widely used to study count data in medicine, economics, and social sciences. This model assumes the response variable to have a Poisson distribution, and also that the logarithm of its expected value can be modeled by a linear combination of unknown parameters. In the Poisson regression model, the response variable for observation i (with $i = 1, \ldots, n$), y_i is modeled as a Poisson random variate with mean μ_i that is specified as a function of a vector of predictor variables $\boldsymbol{X}$ and a vector of parameters $\boldsymbol{\beta}$. The parameter vector $\boldsymbol{\beta}$ is unknown, and we wish to estimate it or test hypotheses about it. These can be done by using the maximum likelihood method and the likelihood ratio test. It is well documented in the reviewed literature that pretest and shrinkage estimators of the James and Stein type have superior performance in terms of asymptotic bias and risk over other estimators considered, under a variety of conditions. Sapra (2003) developed the pretest estimation method for a Poisson regression model. Hossain and Ahmed (2012) extends the shrinkage estimation method for Poisson regression by combining ideas from the recent literature on sparsity patterns. Variable selection is fundamental in statistical modeling. Initially, there may be many variables to consider as candidates for predictors in the model. Some of these variables may not be active and should therefore be excluded from the final model so as to achieve the goal of good prediction accuracy. Researchers are often interested in finding an active subset of predictors that represent a sparsity pattern in the predictor space. In the next step, they may consider this information and use it either in the full model or in the submodel. We follow this procedure in this chapter which is inspired by Stein's result that in a dimension greater than two, efficient estimators can be obtained by shrinking full model estimators in the direction of submodel estimators. In this chapter, we consider the problem of estimating the parameters of Poisson regression model for the purpose of predicting a response variable that may be affected by several potential predictor variables, some of which may be inactive. The prior information from the inactive variables may be incorporated into the estimation procedure to obtain shrinkage estimators. The existing literature shows that the shrinkage estimators significantly improve upon the classical estimators.

We reappraise the properties of shrinkage estimators for the Poisson regression model when it is suspected that some of the parameters may be restricted to a subspace. We develop the properties of shrinkage estimators using the notion of asymptotic distributional bias and risk. The shrinkage estimators are shown to have higher efficiency than the classical estimators for a wide class of models. Furthermore, we consider three different penalty estimators—LASSO, adaptive LASSO, and SCAD—and compare the relative performance with the shrinkage estimators. Monte Carlo simulations reveal that the shrinkage strategy is competitive compared with LASSO, adaptive LASSO, and SCAD when, and only when, there is a moderate or large number of inactive predictors in the model. The shrinkage and penalty strategies are applied to two real data sets to illustrate the usefulness of the procedures in practice. A number of studies have been conducted using the application of shrinkage estimation by several authors: Ahmed et al. (2007), Ahmed et al. (2006), Judge and Mittelhammaer (2004), Ahmed and Saleh (1999), and Ahmed et al. (2012). They developed shrinkage estimation strategies for parametric, semiparametric, and nonparametric linear models.

Thus, the goal of this chapter is to analyze some of the issues involved in parameter estimation for a Poisson regression model when a candidate submodel is available. For example, in genomics research, it is a common practice to test a subset of genetic markers for association with a disease. If the subset is found in a certain population after doing genome-wide association studies, then the subset is tested for disease association in a new population. In this new population, it is possible that genetic markers may be discovered that cannot be found in the first population associated with the disease. Another example can be found in Cameron and Trivedi (1998), who observed that the number of visits to a doctor may be related to sex, age, income, illness, number of reduced activity days, general health questionnaire scores, number of chronic conditions, and dummy variables for two levels (levyplus and freerepa) of health insurance coverage. In the case when prior information is not available, the shrinkage estimation strategy uses a two-step approach. In the first step, a set of covariates (the number of reduced activity days, illness, health questionnaire scores, age, sex, and levyplus) are selected based on the best subset selection procedure and traditional model selection criteria, such as AIC and BIC. The effects of other covariates may be inactive. We then use these inactive variables or linear combinations of them to create a linear subspace of the full parameter space for β. The statistical objective of this chapter to provide a unified estimation strategy which implements both shrinkage and penalty methods for estimating the parameters.

The rest of the chapter is organized as follows. The model and suggested estimators are introduced in Sect. 6.2; the asymptotic properties of the proposed estimators and their asymptotic distributional biases and risks are presented in Sect. 6.3; the results of a simulation study that includes a comparison with three penalty methods are given in Sect. 6.4; application to a real data set and a comparison of listed estimation strategies are described in Sect. 6.5; and Sect. 6.6 contains the concluding remarks.

6.2 Estimation Strategies

Suppose that y_i, given the vector of predictors $\boldsymbol{x}_i$, is independently distributed as Poisson probability distribution, that is

$$f(y_i|\boldsymbol{x}_i) = \frac{e^{-\mu_i}\mu_i^{y_i}}{y_i!}, \quad y_i = 0, 1, 2, \ldots, \quad i = 1, 2, 3, \ldots, n. \tag{6.1}$$

then the mean parameter is

$$\mu_i = \exp(\boldsymbol{x}_i'\boldsymbol{\beta}). \tag{6.2}$$

Here, $\boldsymbol{x}_i = (x_{i1}, x_{i2}, \ldots, x_{ip})'$ is a $p \times 1$ vector of predictors and $\boldsymbol{\beta}$ is a $p \times 1$ vector of regression parameters.

Thus, the log-likelihood function is given by

$$l(\boldsymbol{\beta}) = \sum_{i=1}^{n} \left[y_i\boldsymbol{x}_i'\boldsymbol{\beta} - \exp(\boldsymbol{x}_i'\boldsymbol{\beta}) - \ln(y_i!) \right]. \tag{6.3}$$

The derivatives of the log-likelihood with respect to $\boldsymbol{\beta}$ can be obtained as

$$\frac{\partial l}{\partial \boldsymbol{\beta}} = \sum_{i=1}^{n} \left[y_i - \exp(\boldsymbol{x}_i'\boldsymbol{\beta}) \right] \boldsymbol{x}_i = 0. \tag{6.4}$$

The estimator based on full model or unrestricted maximum likelihood estimator (UE) $\hat{\boldsymbol{\beta}}^{\mathrm{UE}}$ of $\boldsymbol{\beta}$ is obtained by solving the score Eq. (6.4). Clearly, these equations are nonlinear in parameter $\boldsymbol{\beta}$. These can be solved by using an iterative algorithm such as the Newton–Raphson method. Under usual regularity conditions (Santos and Neves 2008) $\hat{\boldsymbol{\beta}}^{\mathrm{UE}}$ is a consistent estimator of $\boldsymbol{\beta}$. Furthermore, it follows p-variate normal distribution as $n \to \infty$ with the variance-covariance matrix $(\boldsymbol{I}(\boldsymbol{\beta}))^{-1}$, where $\boldsymbol{I}(\boldsymbol{\beta}) = \sum_{i=1}^{n} e^{\boldsymbol{x}_i'\boldsymbol{\beta}} \boldsymbol{x}_i \boldsymbol{x}_i'$.

We consider a linear subspace where the unknown p-dimensional parameter vector $\boldsymbol{\beta}$ satisfies a set of p_2 linear restrictions

$$\boldsymbol{H}\boldsymbol{\beta} = \boldsymbol{h}, \tag{6.5}$$

where $\boldsymbol{H}$ is $p_2 \times p$ matrix of rank $p_2 \le p$, and $\boldsymbol{h}$ is a given $p_2 \times 1$ vector of constants. Because $\boldsymbol{H}$ has rank p_2, the p_2 equations may not contain any redundant information about $\boldsymbol{\beta}$.

The submodel or restricted maximum likelihood estimator (RE), $\boldsymbol{\beta}^{\mathrm{RE}}$ of $\boldsymbol{\beta}$ is obtained by maximizing the log-likelihood function (6.3) under the linear restrictions $\boldsymbol{H}\boldsymbol{\beta} = \boldsymbol{h}$.

6.2.1 Shrinkage Estimation Strategies

To start the process, let $l(\hat{\boldsymbol{\beta}}^{\text{UE}})$ and $l(\hat{\boldsymbol{\beta}}^{\text{RE}})$ be the values of the log-likelihood at the full model and submodel estimates respectively, then

$$\begin{aligned} D &= 2[l(\hat{\boldsymbol{\beta}}^{\text{UE}}; y_1, \ldots, y_n) - l(\hat{\boldsymbol{\beta}}^{\text{RE}}; y_1, \ldots, y_n)], \\ &= (\boldsymbol{H}\hat{\boldsymbol{\beta}}^{\text{UE}} - \boldsymbol{h})'[\boldsymbol{H}(\boldsymbol{I}(\boldsymbol{\beta}))^{-1}\boldsymbol{H}']^{-1}(\boldsymbol{H}\boldsymbol{\beta}^{\text{UE}} - \boldsymbol{h}) + o_p(1). \end{aligned}$$

Noting that if the submodel is true then distribution of D converges to χ^2 with p_2 degrees of freedom as $n \to \infty$.

Now we define the shrinkage estimator as

$$\hat{\boldsymbol{\beta}}^{\text{S}} = \hat{\boldsymbol{\beta}}^{\text{RE}} + \left(1 - (p_2 - 2)D^{-1}\right)(\hat{\boldsymbol{\beta}}^{\text{UE}} - \hat{\boldsymbol{\beta}}^{\text{RE}}), \quad p_2 \geq 3.$$

The *shrinkage estimator* (SE) combines the information from the full model and submodel and yields an efficient estimator and subsequently improves the prediction accuracy. Ahmed (1997) among others, shows that the shrinkage estimator is relatively more efficient than the full model-based least square estimator in the classical regression models. Recall that by construction the shrinkage estimator is not a convex combination of the full model and submodel estimators. It is possible that the shrinkage estimator may have the opposite sign of estimator based on the full model. To alleviate this annoying feature of the shrinkage estimator, we suggest using the positive version of this estimator, which is known as positive-part shrinkage estimator (PSE). Let us denote where $z^+ = max(0, z)$, then PSE is defined as

$$\hat{\boldsymbol{\beta}}^{\text{S+}} = \hat{\boldsymbol{\beta}}^{\text{RE}} + \left(1 - (p_2 - 2)D^{-1}\right)^+ (\hat{\boldsymbol{\beta}}^{\text{UE}} - \hat{\boldsymbol{\beta}}^{\text{RE}}).$$

6.2.2 Penalty Estimation

As a family of penalized least squares methods, Park and Hastie (2007) proposed the LASSO version for the Poisson regression model. In a sense, it is a useful method for simultaneous variable selection and estimation of parameters in the selected submodel. This procedure calculates the regression coefficients that minimize the negative log-likelihood function subject to an L_1 penalty on the regression parameter vector.

In this chapter we consider three commonly used penalty estimation strategies.

LASSO

The LASSO estimate of $\boldsymbol{\beta}$ is obtained by minimizing the following function. More specifically,

$$\hat{\boldsymbol{\beta}}^{\mathrm{LASSO}} = \underset{\boldsymbol{\beta}}{\operatorname{argmin}}\{-l(\boldsymbol{\beta}) + \lambda||\boldsymbol{\beta}||_1\}$$
$$= \underset{\boldsymbol{\beta}}{\operatorname{argmin}}\left[-\sum_{i=1}^{n}\left[y_i\boldsymbol{x}_i'\boldsymbol{\beta} - \exp(\boldsymbol{x}_i'\boldsymbol{\beta}) - \ln y_i!\right] + \lambda||\boldsymbol{\beta}||_1\right],$$

where λ is called the tuning parameter, and *argmin* gives the value of $\boldsymbol{\beta}$ which minimizes the function of interest. For large values of λ, this technique produces shrunken estimates of $\boldsymbol{\beta}$, often with many components equal to zero. Park and Hastie (2007) suggested an algorithm that implements the predictor-corrector method to determine the entire path of the coefficient estimates as λ varies from 0 to ∞. The algorithm computes a series of solutions and estimates the coefficients with a smaller λ each time based on the previous estimate.

Adaptive LASSO

The adaptive LASSO is the solution of

$$\hat{\boldsymbol{\beta}}^{\mathrm{aLASSO}} = \underset{\boldsymbol{\beta}}{\operatorname{argmin}}\left[-\sum_{i=1}^{n}\left[y_i\boldsymbol{x}_i'\boldsymbol{\beta} - \exp(\boldsymbol{x}_i'\boldsymbol{\beta}) - \ln y_i!\right] + \lambda\sum_{i=1}^{p}|\beta_i|w_i\right],$$

where w_i's are adaptive weights defined as $w_i = |\hat{\boldsymbol{\beta}}_i|^{-\tau}$ for some positive τ, and $\hat{\beta}_i$ is the maximizer of the log likelihood $l(\boldsymbol{\beta})$. The intuition of the adaptive LASSO is to put large weights to nuisance variables, and gives small weights to active variables, as well as shrink their associated coefficients slightly. In passing, we would like to remark here that adaptive LASSO enjoys oracle properties (Fan and Li 2001) that LASSO does not have.

SCAD

Fan and Li (2001) proposed the smoothly clipped absolute deviation (SCAD) method for linear and generalized linear models. This procedure selects important variables and estimate the regression parameters $\boldsymbol{\beta}$ simultaneously by maximizing the following penalized likelihood function:

$$\hat{\boldsymbol{\beta}}^{\text{SCAD}} = \underset{\boldsymbol{\beta}}{\operatorname{argmin}} \left[-\sum_{i=1}^{n} \left[y_i \boldsymbol{x}_i' \boldsymbol{\beta} - \exp(\boldsymbol{x}_i' \boldsymbol{\beta}) - \ln y_i! \right] + \lambda \sum_{i=1}^{p} p_\lambda(|\beta_i|) \right],$$

where $p_\lambda(\cdot)$ is the smoothly clipped absolute deviation penalty with a tuning parameter λ. The tuning parameter λ is selected using a cross-validation technique. The penalty $p_\lambda(\cdot)$ satisfies $p_\lambda(0) = 0$, and its first-order derivative

$$p_\lambda'(\theta) = \lambda \left[I(\theta \leq \lambda) + \frac{(a\lambda - \theta)_+}{(a-1)\lambda} I(\theta > \lambda) \right],$$

where a is some constant usually taken to be $a = 3.7$ (Fan and Li 2001), and $(t)_+ = tI\{t > 0\}$ is the hinge loss function. The procedure selects some relatively inactive variables by producing zero solutions for their associated regression coefficients.

It is worth noting that the output of the above three penalty methods can be viewed as a shrinkage technique by both shrinking and deleting coefficients. More importantly, it is different from the classical shrinkage estimation strategy in that it weighs all the predicting variables coefficients equally. Penalty estimation procedures do not require a specified linear subspace restriction $\boldsymbol{H\beta} = \boldsymbol{h}$.

Now, we turn our attention to the estimation of regression parameter using a classical approach and provide some asymptotic results in the following section.

6.3 Asymptotic Analysis

We present the expressions for the asymptotic bias and risk of the full model, submodel, and shrinkage estimators.

We will provide these expressions under local alternatives to give a fair analysis. To this end, let $\boldsymbol{\delta} = (\delta_1, \delta_2, \ldots, \delta_{p_2}) \in \Re^{p_2}$ and consider the following sequence of local alternatives:

$$K_{(n)} : \boldsymbol{H\beta} = \boldsymbol{h} + \frac{\boldsymbol{\delta}}{\sqrt{n}}. \tag{6.6}$$

Now, we present two central key results to the study of statistical properties of shrinkage estimators in the following theorems.

Theorem 6.1 *Under the local alternatives $K_{(n)}$ in (6.6) and the usual regularity conditions, as $n \to \infty$,*

1. *$\sqrt{n}(\boldsymbol{H}\hat{\boldsymbol{\beta}}^{UE} - \boldsymbol{h}) \xrightarrow{d} N(\boldsymbol{\delta}, \boldsymbol{HB}^{-1}\boldsymbol{H}')$, where $\boldsymbol{B}_{p \times p} = \lim_{n\to\infty} \frac{I(\boldsymbol{\beta})}{n}$ is nonsingular,*
2. *The quantity D converges to a noncentral chi-squared distribution $\chi^2_{p_2}(\Delta)$ with p_2 degrees of freedom and noncentrality parameter $\Delta = \boldsymbol{\delta}'(\boldsymbol{HB}^{-1}\boldsymbol{H}')^{-1}\boldsymbol{\delta}$.*

Using this theorem, we can obtain the main results of this section. We present (without derivation) the bias and risk expressions for the estimators based on likelihood function.

Suppose that $\boldsymbol{\beta}^*$ is any estimator of $\boldsymbol{\beta}$ and $\boldsymbol{Q}$ is a positive semi-definite matrix, then the quadratic loss function is

$$\mathscr{L}(\boldsymbol{\beta}^*; \boldsymbol{Q}) = \left[\sqrt{n}(\boldsymbol{\beta}^* - \boldsymbol{\beta})\right]^{\top} \boldsymbol{Q} \left[\sqrt{n}(\boldsymbol{\beta}^* - \boldsymbol{\beta})\right]. \qquad (6.7)$$

For theoretical results, we use a general weight $\mathbf{Q}$ on the asymptotic variances and covariances of the estimators. A common choice of $\mathbf{Q}$ is the identity matrix. This is what we use in the simulation study.

The asymptotic distribution function of $\boldsymbol{\beta}^*$ under $K_{(n)}$ is given by

$$G(\boldsymbol{y}) = \lim_{n\to\infty} P\left[\sqrt{n}(\boldsymbol{\beta}^* - \boldsymbol{\beta}) \leq \boldsymbol{y}|K_{(n)}\right],$$

where $G(\boldsymbol{y})$ is a nondegenerate distribution function. We define the *asymptotic distributional risk* (ADR) by

$$\begin{aligned} R(\boldsymbol{\beta}^*; \boldsymbol{Q}) &= \int \cdots \int \boldsymbol{y}' \boldsymbol{Q} \boldsymbol{y} dG(\boldsymbol{y}), \\ &= \text{trace}(\boldsymbol{Q}\boldsymbol{Q}^*), \end{aligned}$$

where $\boldsymbol{Q}^* = \int \cdots \int \boldsymbol{y}\boldsymbol{y}' dG(\boldsymbol{y})$ is the dispersion matrix of $G(\boldsymbol{y})$.

Note that under fixed or nonlocal alternatives, all the estimators are asymptotically equivalent to $\hat{\boldsymbol{\beta}}^{\text{UE}}$, while $\hat{\boldsymbol{\beta}}^{\text{RE}}$ may have an unbounded risk. In order to make an interesting and meaningful comparison and to obtain a nondegenerate asymptotic distribution $G(\boldsymbol{y})$, we will use the local alternatives in (6.6).

The *asymptotic distributional bias* (ADB) of an estimator $\boldsymbol{\beta}^*$ is defined as

$$\text{ADB}(\boldsymbol{\beta}^*) = \lim_{n\to\infty} E\left\{n^{\frac{1}{2}}(\boldsymbol{\beta}^* - \boldsymbol{\beta})\right\} = \int \cdots \int \boldsymbol{y} dG(\boldsymbol{y}),$$

noting that second equality can be established under the model at hand assumptions.

Theorem 6.2 *Under the local alternatives $K_{(n)}$ and the condition of Theorem 6.1, the ADBs of the estimators are*

$$\begin{aligned} ADB(\hat{\boldsymbol{\beta}}^{UE}) &= \mathbf{0}, \\ ADB(\hat{\boldsymbol{\beta}}^{RE}) &= -\boldsymbol{J}\boldsymbol{\delta}, \quad \boldsymbol{J} = \boldsymbol{B}^{-1}\boldsymbol{H}'[\boldsymbol{H}\boldsymbol{B}^{-1}\boldsymbol{H}']^{-1}, \\ ADB(\hat{\boldsymbol{\beta}}^{S}) &= -(p_2 - 2)\boldsymbol{J}\boldsymbol{\delta} E(\chi^{-2}_{p_2+2}(\Delta)), \\ ADB(\hat{\boldsymbol{\beta}}^{S+}) &= -(p_2 - 2)\boldsymbol{J}\boldsymbol{\delta}\left[E(\chi^{-2}_{p_2+2}(\Delta)) - E(\chi^{-2}_{p_2+2}(\Delta) I(\chi^{2}_{p_2+2}(\Delta) < (p_2 - 2)))\right] \\ &\quad -\boldsymbol{J}\boldsymbol{\delta}\Psi_{p_2+2}(p_2 - 2, \Delta), \end{aligned}$$

where $\Psi_\nu(p_2-2,\Delta)$ *is the distribution function of the* $\chi^2_{p_2}(\Delta)$ *distribution and the remaining discussions follows.*

Theorem 6.3 *Under the local alternatives* $K_{(n)}$ *and the assumptions of Theorem 6.1, the risk of the estimators are*

$$\begin{aligned}
R(\hat{\boldsymbol{\beta}}^{UE};\boldsymbol{Q}) &= trace[\boldsymbol{QB}^{-1}],\\
R(\hat{\boldsymbol{\beta}}^{RE};\boldsymbol{Q}) &= R(\hat{\boldsymbol{\beta}}^{UE};\boldsymbol{Q}) - trace[\boldsymbol{QJHB}^{-1}] + \boldsymbol{\delta}'(\boldsymbol{J}'\boldsymbol{QJ})\boldsymbol{\delta},\\
R(\hat{\boldsymbol{\beta}}^{S};\boldsymbol{Q}) &= R(\hat{\boldsymbol{\beta}}^{UE};\boldsymbol{Q}) - 2(p_2-2)trace[\boldsymbol{QJHB}^{-1}]\{2E(\chi^{-2}_{p_2+2}(\Delta))\\
&\quad -(p_2-2)E(\chi^{-4}_{p_2+2}(\Delta))\} + (p_2-2)\boldsymbol{\delta}'(\boldsymbol{J}'\boldsymbol{QJ})\boldsymbol{\delta}\{2E(\chi^{-2}_{p_2+2}(\Delta))\\
&\quad -2E(\chi^{-4}_{p_2+2}(\Delta)) + (p_2-2)E(\chi^{-4}_{p_2+4}(\Delta))\},\\
R(\hat{\boldsymbol{\beta}}^{S+};\boldsymbol{Q}) &= R(\hat{\boldsymbol{\beta}}^{S};\boldsymbol{Q}) - \boldsymbol{\delta}'(\boldsymbol{J}'\boldsymbol{QJ})\boldsymbol{\delta}E[(1-(p_2-2)\chi^{-2}_{p_2+4}(\Delta))^2 I(\chi^2_{p_2+4}(\Delta)<p_2-2)]\\
&\quad - trace[\boldsymbol{QJHB}^{-1}]E[(1-(p_2-2)\chi^{-2}_{p_2+2}(\Delta))^2 I(\chi^2_{p_2+4}(\Delta)<p_2-2)]\\
&\quad + 2\boldsymbol{\delta}'(\boldsymbol{J}'\boldsymbol{QJ})\boldsymbol{\delta}E[(1-(p_2-2)\chi^{-2}_{p_2+4}(\Delta))I(\chi^2_{p_2+4}(\Delta)<p_2-2)].
\end{aligned}$$

Based on Theorem 6.2, for any $\boldsymbol{Q}\in\boldsymbol{Q}^D$ and all $\boldsymbol{\delta}$ and under $\{K_n\}$,

$$R(\hat{\boldsymbol{\beta}}^{S+};\boldsymbol{Q}) \le R(\hat{\boldsymbol{\beta}}^{S};\boldsymbol{Q}) \le R(\hat{\boldsymbol{\beta}}^{UE};\boldsymbol{Q})$$

where

$$\boldsymbol{Q}^D = \left\{\boldsymbol{Q}: \frac{\text{trace}[\boldsymbol{QJHB}^{-1}]}{Ch_{max}[\boldsymbol{QJHB}^{-1}]} \ge \frac{p_2+2}{2}\right\},$$

and $Ch_{max}(\cdot)$ is the maximum characteristic root. When $\boldsymbol{\Delta}=0$, the following relationship holds:

$$R(\hat{\boldsymbol{\beta}}^{RE};\boldsymbol{Q}) < R(\hat{\boldsymbol{\beta}}^{S+};\boldsymbol{Q}) < R(\hat{\boldsymbol{\beta}}^{S};\boldsymbol{Q}) < R(\hat{\boldsymbol{\beta}}^{UE};\boldsymbol{Q}).$$

However, for small values of $\Delta(>0)$,

$$R(\hat{\boldsymbol{\beta}}^{S+};\boldsymbol{Q}) < R(\hat{\boldsymbol{\beta}}^{S};\boldsymbol{Q}) < R(\hat{\boldsymbol{\beta}}^{UE};\boldsymbol{Q}) < R(\hat{\boldsymbol{\beta}}^{RE};\boldsymbol{Q}).$$

More importantly, $\boldsymbol{\beta}^{S+}$ dominates the $\boldsymbol{\beta}^{UE}$ in the entire parameter space induced by Δ.

In an effort to numerically appraise the performance of suggested estimators, we conduct a simulation study to compare the performance of the likelihood-based estimators and the penalty estimators for selected sample sizes.

For relative comparison, we consider the full model or unrestricted estimator $\hat{\boldsymbol{\beta}}^{UE}$ as the "baseline" estimator. Hence, the performance of the other estimators appraised in terms of the simulated MSE relative to the MSE of $\hat{\boldsymbol{\beta}}^{UE}$ (RMSE). For any estimator $\hat{\boldsymbol{\beta}}^{\star}$, the simulated relative MSE (RMSE) of $\hat{\boldsymbol{\beta}}^{\star}$ to $\hat{\boldsymbol{\beta}}^{UE}$ is given by

$$\text{RMSE}(\hat{\boldsymbol{\beta}}^*) = \frac{\text{simulated MSE}(\hat{\boldsymbol{\beta}}^{\text{UE}})}{\text{simulated MSE}(\hat{\boldsymbol{\beta}}^*)}.$$

Evidently, an RMSE larger than one indicates the degree of superiority of the estimator $\hat{\boldsymbol{\beta}}^*$ over $\hat{\boldsymbol{\beta}}^{\text{UE}}$.

6.4 Monte Carlo Simulation

We use Monte Carlo simulation experiments to assess the relative risk performance of the listed estimators. We consider the following model for data generation

$$\log y_i = \boldsymbol{x}_i'\boldsymbol{\beta}.$$

We simulate $\boldsymbol{x}_i{}' = (x_{i1}, x_{i2}, \ldots, x_{in})$ from a multivariate standard normal distribution.

We consider the partition of $\beta = (\boldsymbol{\beta}_1', \boldsymbol{\beta}_2')'$. The coefficients $\boldsymbol{\beta}_1$ and $\boldsymbol{\beta}_2$ are $p_1 \times 1$ and $p_2 \times 1$ vectors, respectively, with $p = p_1 + p_2$. We assume that the model is sparse and set $\boldsymbol{\beta}_2 = \mathbf{0}$. Thus, the interest is in estimating $\boldsymbol{\beta}_1$. However, the assumption of sparsity may not be a realistic one. We consider the true value of $\boldsymbol{\beta}$ at $\boldsymbol{\beta} = (\boldsymbol{\beta}_1', \mathbf{0})'$ as $\boldsymbol{\beta}_1 = (0.2, -1.2, 0.1, 0.2)$ for simulating the data. In Table 6.1, we present relative MSEs of submodel, shrinkage, and penalty estimators (LASSO, adaptive LASSO, and SCAD) with respect to the full model estimator for $n = 50$ with $p_1 = 4$. The simulation results are summarized in Table 6.1 for $\Delta = 0$. We estimated the tuning parameter λ for the penalty estimators using a 10-fold cross-validation.

Table 6.1 is computed based on the assumption that the selected submodes is correct, that is, $\boldsymbol{\beta}_2 = \mathbf{0}$. However, this assumption maybe hard to justify in some real situations. In an effort to provide a meaningful comparison, we define the parameter $\Delta^* = ||\boldsymbol{\beta} - \boldsymbol{\beta}^{(0)}||^2$, where $\boldsymbol{\beta}^{(0)} = (\boldsymbol{\beta}_1', \mathbf{0})'$ and $||\cdot||$ is the Euclidian norm. Further data were generated such that Δ is between 0 and 2. The RMSE of the submodel and shrinkage estimators to the full model estimator is computed for $n = 50$ and $p_2 = 10$. The penalty estimators were not considered for $\Delta > 0$, because by design these methods select a submodel and estimate the parameters of the selected submodel, under the assumption that the selected submodel is the best one and no further investigation is required. On the contrary, the shrinkage estimators treat this situation well by adapting to the $\boldsymbol{\beta}_2 \neq \mathbf{0}$ case. Thus, we examine the performance of the shrinkage estimators and study how they compare to submodel and full model when p is fixed. The results for a sample size of $n = 50$ are presented in Fig. 6.1 and Table 6.2, and we obtain the following analysis.

Table 6.1 reveals that the RMSE of all the estimators increases as the number of p_2 increases. Moreover, as we would expect, the submodel estimator is the best among the class of estimators. Furthermore, all the estimators are superior to the full model estimators. Table 6.1 indicates that the penalty estimation strategy performs better

Table 6.1 RMSE of the listed estimators to the full model estimator assuming submodel is correct

Method	$p_2 = 3$	$p_2 = 5$	$p_2 = 7$	$p_2 = 9$	$p_2 = 11$	$p_2 = 13$	$p_2 = 15$
RE	2.46	3.45	5.48	7.37	9.54	13.82	17.56
SE	1.25	1.58	2.29	3.03	3.64	4.38	4.99
PSE	1.36	1.63	2.82	3.68	4.41	5.44	6.15
LASSO	1.45	1.69	1.92	2.46	2.85	3.88	4.55
ALASSO	1.52	1.78	2.04	2.38	2.93	3.96	4.78
SCAD	1.56	1.89	2.35	2.98	4.16	5.28	5.79

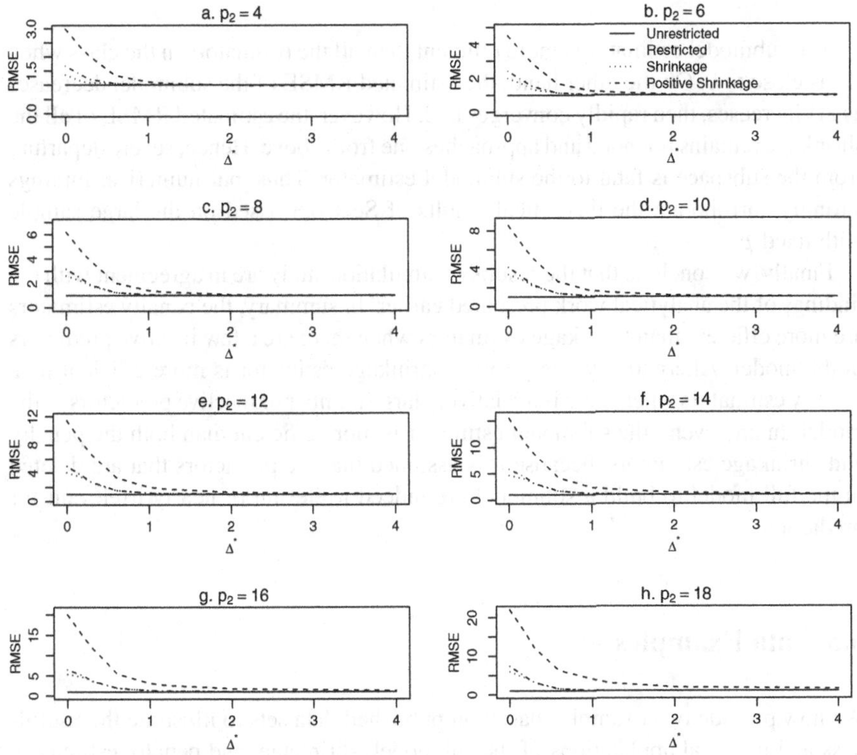

Fig. 6.1 RMSE of the proposed estimators to the full model estimator with $p_1 = 4$, $p_2 = 4, 6, 8, 12, 14, 16, 18$, and $n = 60$

than the shrinkage strategy when, and only when, the number of p_2 in the model is relative small. We see from Table 6.1 that, when $(p_1, p_2) = (4, 5)$, the penalty estimators are more efficient than the positive shrinkage estimator. On the other hand, the positive shrinkage estimator is more efficient than estimators for larger values of p_2. This is an interesting and surprising observation. Hence, we suggest to use the positive shrinkage estimator when the number of inactive predictors is relatively large.

Table 6.2 RMSE of the proposed estimators to the full model estimator with $n = 50$ and $p_2 = 10$

Δ^*	$\hat{\beta}^{RE}$	$\hat{\beta}^{S}$	$\hat{\beta}^{S+}$
0.00	3.52	2.16	2.45
0.10	3.22	1.92	2.12
0.20	2.56	1.64	1.68
0.30	2.06	1.36	1.37
0.60	1.42	1.11	1.11
1.00	1.19	1.06	1.06
1.20	1.13	1.03	1.03
2.00	1.00	1.00	1.00

The submodel estimator is more efficient than all the estimators in the class when Δ^* is close to 0. On the other hand, the estimated RMSE of the submodel decreases as Δ^* increases, then rapidly converges to 0. However, the estimated RMSE of all the shrinkage remains bounded and approaches one from above. Hence, severe departure from the subspace is fatal to the submodel estimator. Thus, our numerical findings strongly corroborate the theoretical results of Sect. 6.3 based on the large sample with fixed p.

Finally, we conclude that the results of simulation study are in agreement with the findings of the analytical work presented earlier. In summary, the penalty estimators are more efficient than shrinkage estimators when there are a few inactive predictors in the model. Alternatively, the positive shrinkage estimator is more efficient than penalty estimators when there is a relatively large number of inactive predictors in the model. In any event, the submodel estimator is more efficient than both the penalty and shrinkage estimators, because it is assumed that the predictors that are deleted in the full model to build a submodel are indeed irrelevant or nearly irrelevant for prediction.

6.5 Data Examples

We now provide two examples based on published data sets to illustrate the usefulness and practical applications of the submodel, shrinkage, and penalty estimation strategies.

Australian Health Survey Data

We refer to Cameron and Trivedi (1998) for a detailed description of the data. A total of 5,190 individuals over 18 years of age answered all of the essential questions that are recorded in this data set. The main objective of this survey was to study the relationship between the number of consultations with a doctor and the type of health insurance, health status, and socioeconomic indicators. The response variable

Table 6.3 RMSE of the proposed estimators to the full model estimator

Estimators	UE	RE	SE	PSE	LASSO	Adaptive LASSO	SCAD
RMSE	1.00	1.90	1.06	1.07	1.11	1.14	1.17

of interest is the number of visits to a doctor that were made during a 2 week interval, and the covariates of interest are sex, age, income, illness, number of reduced activity days, general health questionnaire scores, number of chronic conditions, and dummy variables for two levels (levyplus = 1 if respondent is covered by private health insurance, 0 otherwise and freerepa = 1 if respondent is covered free by government, 0 otherwise) of health insurance coverage.

The preliminary analysis-based maximum likelihood estimation categories suggests that reduced activity days (β_1), illness (β_2), health questionnaire scores (β_3), age (β_4), sex (β_5), levyplus (β_6) as the useful variables, while the rest of the variables income (β_7), number of chronic conditions (β_8), and freerepa (β_9) are not important variables for predicting the number of visits to a doctor. We may use this information as an auxiliary information to form subspace or to obtain a submodel. Alternatively, one can use the existing variable selection procedures to obtain a submodel. In any event, for this example, we set $\boldsymbol{\beta}_2 = (\beta_7, \beta_8, \beta_9) = (0, 0, 0)$, $p = 9$, $p_1 = 6$, $p_2 = 3$. The MSE using bootstrap resampling of size 1,000 are computed. The relative MSE is reported in Table 6.3. The table reveals that the penalty estimators are efficient more than the shrinkage estimators. Not surprisingly, the submodel estimator is the most efficient estimator as compared to the class of estimators studied here.

Takeover Bids Data

In this example, we apply our estimation strategies to a takeover bids data set provided by Cameron and Trivedi (1998). The data set includes the number of bids received by 126 U.S. firms that were targets of tender offers during the period of 1978–1985, and were taken over within 52 weeks of the initial offer. The response count variable is the number of bids (numbids) after the initial bid received by the target firm. The data are based on eight explanatory variables.

The initial maximum likelihood inference leads us to conclude that the bid price (β_1), management invitation (β_2), and total book value of assets on the takeover bids (β_3) may be useful variables for prediction purposes. However, other variables in the initial model, percentage of stock held by institutions (β_4), legal defense by lawsuit (β_5), proposed changes in asset structure (β_6), proposed changes in ownership structure (β_7), and government intervention (β_8) are not significant to predict the number of takeover bids received by targeted firms. We wish to treat this information as auxiliary information to improve the estimation accuracy of the remaining three

Table 6.4 RMSE of the proposed estimators to the full model estimator

Estimators	UE	RE	SE	PSE	LASSO	Adaptive LASSO	SCAD
RMSE	1.00	2.27	1.22	1.24	1.51	1.56	1.74

parameters in the model, resulting in better prediction performance. In this example, we set $\boldsymbol{\beta}_2 = (\beta_4, \beta_5, \beta_6, \beta_7, \beta_8) = (0, 0, 0, 0, 0)$. Thus, we have $p_1 = 3$, $p_2 = 5$ with $n = 126$. We use the auxiliary information to construct submodel and shrinkage estimators. We also apply the penalty estimation procedure and reappraise its performance with other estimators for these data. The results are reported in Table 6.4 and the findings remain the same as those of previous example.

In passing, we would like to remark here that in both data examples the estimators perform better than shrinkage estimators since both models have a relatively small number of parameters and a small number of the predicting variables found to be insignificant in the models. In any event, the conclusions from the data examples are consistent with both analytical and numerical findings.

6.6 Chapter Summary

In this chapter, we consider the estimation problem in a Poisson regression model. We systematically compare the performance of full model, submodel, shrinkage, and penalty estimators when the full model may be sparse. We appraise the risk properties of the submodel and shrinkage estimators both analytically and numerically. However, the risk properties of the penalty estimators were assessed through Monte Carlo simulation, since analytical solution is not available.

The simulation study leads to conclude that the penalty and shrinkage estimation strategies are competitive and provide a good solution when the full model at hand is sparse. The positive shrinkage estimator performs better than penalty estimators when the number of inactive predictors is moderate or relatively large in the model. On the other hand, penalty estimator is more efficient than the shrinkage estimators when the number of regression coefficients close to zero is small. However, it is important to note that penalty estimator is useful when $n < p$. Furthermore, we reconfirm that SCAD and adaptive LASSO perform better than the LASSO estimator. We suggest using either SCAD or adaptive LASSO when the full model is really sparse. However, sparsity is a strong assumption and some thoughts should be given before doing simultaneous variable selection and estimation.

Finally, we used the suggested estimation strategies on two published data sets to investigate the relative performance of all estimators to the maximum likelihood estimator based on the full model. The conclusions drawn based on data set strongly corroborate our analytical and simulated results.

References

Ahmed, S. E. (1997). Asymptotic shrinkage estimation: the regression case. *Applied Statistical Science, II*, 113–139.

Ahmed, S. E., Doksum, K. A., Hossain, S., & You, J. (2007). Shrinkage, pretest and absolute penalty estimators in partially linear models. *Australian and New Zealand Journal of Statistics, 49*, 435–454.

Ahmed, S. E., Hossain, S., & Doksum, K. A. (2012). LASSO and shrinkage estimation in Weibull censored regression models. *Journal of Statistical Planning and Inference, 12*, 1272–1284.

Ahmed, S. E., Hussein, A. A., & Sen, P. K. (2006). Risk comparison of some shrinkage m-estimators in linear models. *Nonparametric Statistics, 18*(4–6), 401–415.

Ahmed, S. E., Saleh, A. K., & Md, E. (1999). Estimation of regression coefficients in an exponential regression model with censored observation. *Journal of Japan Statistical Society, 29*, 55–64.

Cameron, A. C., & Trivedi, P. K. (1998). *Regression Analysis of Count Data*. UK: Cambridge University Press.

Fan, J., & Li, R. (2001). Variable selection via nonconcave penalized likelihood and its oracle properties. *Journal of the American Statistical Association, 96*(456), 1348–1360.

Hossain, S., & Ahmed, S. E. (2012). Shrinkage and penalty estimators in a Poisson regression mode. *Australian and New Zealand Journal of Statistics, 54*, 359–373.

Judge, G. G., & Mittelhammaer, R. C. (2004). A semiparametric basis for combining estimation problem under quadratic loss. *Journal of the American Statistical Association, 99*, 479–487.

Park, M., & Hastie, T. (2007). An L_1 regularization-path algorithm for generalized linear models. *Journal of the Royal Statistical Society: Series B, 69*, 659–677.

Santos, J. A., & Neves, M. M. (2008). A local maximum likelihood estimator for poisson regression. *Metrika, 68*, 257–270.

Sapra, S. K. (2003). Pre-test estimation in poisson regression model. *Applied Economics Letters, 10*, 541–543.

